curriculum
connections

Technology Through the Ages
# The Ages of Steam and Electricity

BROWN BEAR BOOKS

Published by Brown Bear Books Limited

An imprint of:
**The Brown Reference Group Ltd**
68 Topstone Road
Redding
Connecticut 06896
USA
www.brownreference.com

© 2009 The Brown Reference Group Ltd

ISBN: 978-1-933834-86-3

All right reserved. This book is protected by copyright. No part of it may be reproduced, stored in a retrieval system, or transmitted in any form or by any means, without the prior permission in writing of the Publisher, nor be otherwise circulated in any form of binding or cover other than that in which it is published and without a similar condition including this condition being imposed on the subsequent publisher.

Editorial Director: Lindsey Lowe
Managing Editor: Tim Harris
Project Director: Graham Bateman
Editors: Briony Ryles, Derek Hall
Designer: Steve McCurdy
Picture Research: Steve McCurdy

**Library of Congress Cataloging-in-Publication Data available upon request**

**Picture Credits**

Cover Image
Steam train wheels (Shutterstock, Julie Richards)

Shutterstock:
46 Sebastian Kaulitzki; 69 Margo Harrison; 86–87 Nikki Bidgood

Photos.com:
7; 10–11; 16–17; 25; 26; 27; 28; 29; 31; 32–33; 37; 40–41; 43; 45; 49; 53; 58; 63; 67; 70; 71; 72; 77; 78; 81; 83

Artwork © The Brown Reference Group Ltd

The Brown Reference Group Ltd has made every effort to trace copyright holders of the pictures used in this book. Anyone having claims to ownership not identified above is invited to contact The Brown Reference Group Ltd.

Printed in the United States of America

# Contents

| | | | |
|---|---|---|---|
| Introduction | 4–5 | Sources of electricity | 52–55 |
| Water turbines | 6–9 | The invention of radio | 56–61 |
| Iron and steel | 10–15 | Elusive electrons | 62–65 |
| The Great Exhibition | 16–19 | The first automobiles | 66–69 |
| Genetics and Mendel | 20–23 | Airships | 70–75 |
| The electric telegraph | 24–27 | Airplanes | 76–81 |
| Telephones | 28–29 | Capturing sound | 82–85 |
| Submarines | 30–35 | Plastics | 86–89 |
| The Periodic Table | 36–39 | Timelines | 90–105 |
| Mapping the Moon and Mars | 40–43 | Glossary | 106 |
| Germs and disease | 44–47 | Index | 107–112 |
| The internal combustion engine | 48–51 | | |

# Introduction

*Technology through the Ages* forms part of the Curriculum Connections project. This six-volume set describes the story of scientific discovery from the earliest use of fire and the development of the wheel through to space travel, modern computing, and the Human Genome Project. Each volume in the set covers a major historical period, ranging from prehistory up to modern times.

Within each volume there are two types of article:

**In-Depth articles** form the core of the work, and focus on scientific discoveries and technological progress of particular note, giving background to the topic, information about the people involved, and explanations of how the discoveries or inventions have been put to use. Each article focuses on a particular step forward that originated within this period, but the articles often extend back into the history of the topic or forward to later developments to help give further context to each subject. Boxed features add to the information, often explaining scientific principles.

Within each article there are two key aids to learning, which are to be found in color bars located in the margins of each page:

*Curriculum Context* sidebars indicate to the reader that a subject has particular relevance to certain key State and National Science and Technology Education Standards up to Grade 12.

*Glossary* sidebars define key words within the text.

**Timeline articles**, to be found at the end of each volume, list year-by-year scientific discoveries, inventions, technological advances, and key dates of exploration. For each period, the timelines are divided into horizontal bands that each focus on a particular theme of technology or science.

A summary *Glossary* lists the key terms defined in the volume, and the *Index* lists people and major topics covered. Fully captioned *Illustrations* play a major role in the set, and include early prints and paintings, contemporary photographs, artwork reconstructions, and explanatory diagrams.

## About this Volume

In this volume (*The Ages of Steam and Electricity—1825 through 1910*) we cover a period of great innovation and discovery, powered by the burgeoning wealth and demands of growing populations. The mid-19th century was the pinnacle of the steam age, with railroads spreading across Europe and America and steamships crossing the oceans. While the steam engine had generally replaced the ancient waterwheel as a source of power, it received a new lease of life with the invention of the water turbine, which is still employed today in hydroelectric power plants.

A new innovation was the international exhibition, such as the Great Exhibition of 1851 in London, which celebrated and publicized the great advances in science and technology. The electric telegraph, and later the telephone, carried the news across the world.

Toward the end of the 18th century and into the early 19th century, transportation took steps forward. Airships had conquered the skies to a limited extent, but with the arrival of the Wright brothers' heavier-than-air machine, the era of the airplane arrived. On the road, the development of the internal combustion engine heralded the invention of the automobile, and the introduction of mass production methods.

In science, Russian chemist Dmitri Mendeleev brought order to chemistry with his Periodic Table, and Austrian monk Gregor Mendel laid down the fundamental laws of genetics. But it was the discovery of the nature of electricity that had the greatest impact. It took the discovery of the electron by English physicist J. J. Thomson in 1897 to identify the elusive carrier of electric charge. This led to the new science of electronics, beginning with the invention of radio. Radio soon benefited from the invention of electron tubes and the various circuits that employed them.

# Water turbines

The waterwheel was one of the earliest machines to produce power. It continued to be used until the 19th century, when it was gradually eclipsed by the steam engine. But there is another machine that uses water power—the water turbine. In modern hydroelectric plants the water turbine is still used to harness the energy of flowing water to generate electricity.

**Curriculum Context**

Students should be aware that scientists and inventors conduct investigations for a wide variety of reasons.

**Patent**

A grant made by a government that confers upon the creator of an invention the sole right to make, use, and sell that invention for a set period of time.

A waterwheel is an inefficient means of producing power because energy is wasted when water flows over the edges of the paddles. In a water turbine, the paddles, or blades, are encased so that water cannot run off the edges. All the power transfers to the blades. French engineer Claude Burdin (1790–1873) coined the word "turbine" in 1824, and much early development work on water turbines was done in France. In 1826, one of Burdin's students, Benoît Fourneyron (1802–67), set out to gain a prize of 6,000 francs offered to the inventor of a successful turbine. He won it in 1833 with a machine that produced 50 horsepower. This was an outward-flow reaction turbine with a 12-inch (30-cm), 30-bladed wheel that rotated at a rate of over 2,000 revolutions a minute. By 1855, large Fourneyron turbines were generating as much as 800 horsepower.

In about 1820, French engineer Jean-Victor Poncelet (1788–1867) had designed a centrifugal turbine in which the blades were set close to the wheel's axis and the water flowed in at the center. Known as the vortex wheel, the design was patented in the United States in 1838 by Samuel Howd. In England in 1840, Irish-born James Thomson (1822–92) devised a way of controlling the water flow to the turbine blades internally. The American engineer Uriah Boyden (1804–79) made further improvements in 1844. Thomson patented his design in Belfast in 1850. It had a horizontal turbine wheel and was made in England.

A cross section through an early water mill. Water flows downhill into tanks and then past paddles. This causes the paddles to rotate, turning mill wheels above them to which they are connected by a vertical shaft.

## Types of turbines

There are three main types of modern turbines, named after their inventors. Anglo-American engineer James Francis (1815–92) invented the Francis turbine in 1849. This is a reaction turbine with totally encased blades. It has a submerged horizontal wheel with up to 24 curved blades. They are surrounded by an outer set of guide vanes that direct the flowing water against the turbine blades. It works best at medium water pressure. Francis emigrated from England to the United States in 1833, and initially worked on railroad construction. In 1837, he became chief engineer for the locks and canals on the Merrimack River, and designed his turbine to make use of the river's water power.

> **Curriculum Context**
>
> Students should be able to demonstrate the principles of water pressure and dynamics.

### Curriculum Context

Students should understand that people continue inventing new ways of doing things and getting work done.

The Pelton wheel was invented in about 1870 by American engineer Lester Pelton (1829–1908), who made improvements to the waterwheels used to drive machinery in the California gold mines. It is a type of impulse turbine, in which a jet of water from a nozzle strikes bucket-shaped turbine blades and causes the turbine to rotate. It has a vertical turbine wheel on a horizontal axle, and works best with high water pressure. Pelton patented his invention in 1880 and later sold the rights to the Pelton Water Wheel Company of San Francisco. Modern versions of the turbine have efficiencies in the order of 90 percent.

The Kaplan turbine, invented in 1913 by Austrian mechanical engineer Viktor Kaplan (1876–1934), is also a type of reaction turbine, designed to work in slow-flowing water. The turbine wheel has up to eight variable-pitch blades, and resembles a ship's propeller mounted vertically. It is the type most often used today in hydroelectric and tidal power plants.

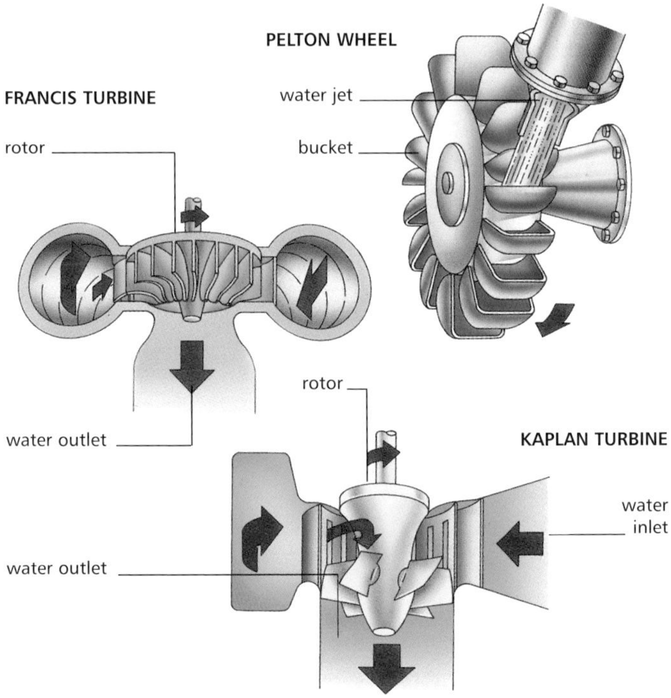

In reaction turbines, such as the Francis turbine and the Kaplan turbine, the direction of water changes after the wheel or rotor has turned. The Pelton wheel is an impulse turbine that spins when water is directed at buckets on the wheel.

**FRANCIS TURBINE**
- rotor
- water outlet

**PELTON WHEEL**
- water jet
- bucket

**KAPLAN TURBINE**
- rotor
- water inlet
- water outlet

# Harnessing the Power

In the United States, a notable example of the way in which the power of the waterwheel fueled the march of industrialization was to be seen at the Boott Cotton Mills in Lowell, Massachusetts. During the early part of the 19th century, a group of Boston merchants saw an opportunity to capitalize on the growing mechanization in the production of textiles, including spinning and weaving. They established the Boston Associates with the aim of integrating the various steps of cotton manufacturing.

One of the associates was Francis Cabot Lowell (1775–1817). He learned about power looms during a visit to England and, together with the other associates, set about building new machinery, and constructing the world's first integrated cotton factory in Waltham, Massachusetts, in 1814. Success followed rapidly, and the men decided to embark on an even more ambitious enterprise—the establishment of an entire town devoted to textile manufacture. They chose a site next to the falls of the Merrimack River, 25 miles (40 km) north of Boston, and called their new town Lowell. A series of canals was dug to harness the energy of the falling water and use it to power a number of mills.

The mills flourished during the 1820s and 1830s, and were a commercial success. By 1835, Lowell had 22 mills. To maximize profits, the owners looked for ways to improve their machinery, including the waterwheels. The wheels that originally powered the mills were capable of converting only 75 percent of the potential energy of water into mechanical energy. The associates encouraged American engineers James Francis (1815–92) and Uriah Boyden (1804–79) to make improvements in the water turbine that had been invented in France in the 1830s. The improved turbines that resulted could convert 90 percent of water's potential energy for use in powering the factory machines. By 1855, Lowell had 55 mills employing more than 13,000 workers, who were producing 2.2 million yards (2 million m) of cotton cloth a week.

Lowell pioneered the female-only workforce. The Lowell Mill "girls" were often daughters of large New England farming families, eager to escape the hardships of farm life, or to earn money independently before getting married. They lived under close supervision in boarding houses owned by the mill company.

Although conditions outside work were good—the company built churches and provided evening classes and activities—work in the mills was hard. A typical working day lasted 12 hours, and the women worked six days a week. Wages were extremely low, and the demand for the company to raise productivity in the face of increasing competition from other textile factories meant that the women were asked to do more work for less money.

Although they fought back by forming a labor union and going on strike, they were eventually defeated by the arrival of large numbers of immigrants fleeing from the Irish potato famine of the 1840s, and willing to work for lower wages. Despite these problems, the mills at Lowell represented a revolution in textile production, and a significant step along the path of American industrialization in the 19th century.

# Iron and steel

Iron has been a valuable material to humankind since prehistory. Once smelting developed, iron could be formed into all manner of items, from cannons to fireplaces. Steel production made possible mighty engineering feats, such as skyscrapers and huge bridges.

Prehistoric people used iron from meteorites, but it was not until around 1500 B.C. that iron smelting was carried out on a large scale. The first people to do so were the Hittites, in what is now Turkey. Iron ore was

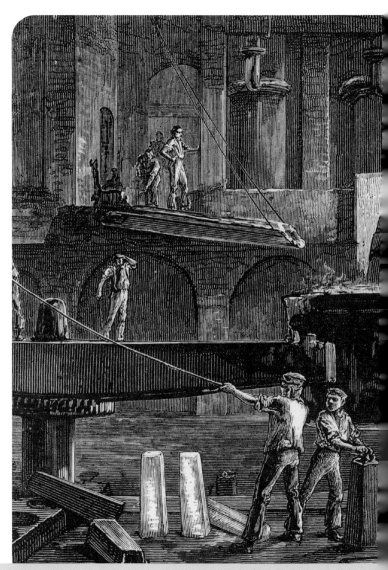

smelted in a charcoal fire, producing metal in a solid lump that was hammered to squeeze out impurities. The final product was wrought iron. For thousands of years, this was the main way iron was produced. The metal was used mainly to make weapons and tools.

## First blast furnaces

Around 600 B.C., the Chinese developed a new, taller kind of furnace in which fuel and crushed iron ore were loaded in at the top and air was forced in at the bottom. This "blast furnace" was hot enough to produce iron in liquid form and in much greater quantities. The

### Wrought iron

A form of iron that is tough, malleable, and relatively soft; it contains usually less than 0.1 percent carbon and 1 or 2 percent of slag.

Steel production in Britain in about the 1880s. In such times, the work was hot and sometimes also dangerous.

Iron and steel

iron produced could be poured into molds to make large objects by casting. Much later, around 1400 A.D, the blast furnace appeared independently in Europe.

By the beginning of the 18th century, charcoal was becoming scarce. In 1709, British ironworker Abraham Darby (1678–1717) first used coke to fuel a blast furnace. Coke was a high-carbon fuel made by heating coal in a closed oven, and its use was an early landmark of the Industrial Revolution. Cast iron increased in importance as a technological material, and in 1779 Darby's descendants built, in Shropshire, England, the first iron bridge outside China. Britain led the way in ironmaking developments until about 1870, when Germany and the United States began to predominate.

### Improving the process

Cast iron is too brittle for many industrial purposes, so wrought iron remained in demand. However, its use was hampered by the small amounts that could be produced. In 1784, a new method called the puddling process was patented by British ironworker Henry Cort (1740–1800). Hot air from a coal fire was directed down onto pig iron from a blast furnace, melting the pig iron and converting it into wrought iron. The process made wrought iron much cheaper. Demand for wrought iron soared as it was used for railroad tracks, bridges, and the new iron ships.

### The advent of steel

In 1856, there was a further development, when British inventor Henry Bessemer (1813–98) devised a process for making steel cheaply and in large quantities. Steel is iron combined with small amounts of carbon. It is strong and tough, but has to be made carefully to control impurities. In ancient India and elsewhere, some steel was made by heating solid wrought iron with charcoal. After hammering, the result was a kind of sandwich of steel, iron, and nonmetallic impurities.

> **Curriculum Context**
>
> Students should be aware that in history, diverse cultures have contributed scientific knowledge and technological inventions.

> **Curriculum Context**
>
> The curriculum expects students to describe how and why technology evolves.

# Blast Furnace

In medieval blast furnaces, water-powered bellows forced air through pipes into the bottom of the furnace, which was loaded with burning charcoal and iron ore. The blast of air allowed far higher temperatures to be reached than in a traditional open-hearth furnace, while the iron ore remained in contact with the charcoal for a much longer time. The ore therefore absorbed more carbon, lowering its melting point sufficiently for molten iron to be produced.

Today, blast furnaces are huge steel cylinders lined with heat-resistant bricks. Giant stoves heat the air (up to 2,100°F, or 1,150°C) that is blasted in at the bottom of the furnace through pipes called tuyères. Wagons take ore, coke, and limestone up a ramp to the top of the furnace. In this type of furnace, a small "bell" is lowered to let the material into the first compartment. The small bell then closes the top before a large bell drops, letting the material fall into the furnace.

The coke burns first, reacting with oxygen in the air to form carbon monoxide gas. This gas acts as a reducing agent by removing oxygen from the iron ore, turning it into metallic iron which, in turn, becomes molten iron in the intense heat. The molten iron is removed regularly through a hole at the bottom of the furnace.

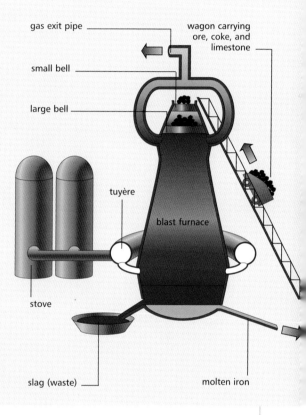

Bessemer's method involved pouring molten iron from the blast furnace into a large vessel, the Bessemer converter. A similar, less successful process was devised in the United States by William Kelly (1811–88).

A second important method, the Siemens-Martin furnace, came into use from the 1860s. The invention of the German brothers William (1823–83) and Friedrich Siemens (1826–1904), this furnace consisted of an open-hearth furnace with brick chambers at either end. Hot air was passed through one of the chambers and over the hearth, which contained pig iron. The hot waste gases passed through the second chamber, heating up the bricks. The direction of the airflow was then reversed, and the hot bricks heated the incoming air. Both these methods were used well into the 20th century. However, they have now been mostly superseded by the basic oxygen furnace, which uses a water-cooled lance to blast pure oxygen at supersonic velocity into a mixture of blast-furnace iron and scrap metal.

## Bessemer Converter

The invention of the Bessemer converter in 1856 made the large-scale production of cheap steel a possibility for the first time in human history. It was based on the discovery by British metallurgist Robert Mushet that the addition of a carbon-manganese alloy to molten iron helped remove oxygen, and alter the carbon content of steel.

The converter consisted of a large, conical steel vessel lined with heat-resistant bricks, and set between pivoting trunnions so that it could be tilted to pour the steel out. Molten pig iron produced by a blast furnace was poured into the top of the converter, while cold air was blasted into the bottom through tuyères in its base. The oxygen in the air combined with impurities in the metal and the carbon-manganese alloy to produce steel. Basic oxygen furnaces are now more commonly used to make steel. They resemble the Bessemer converter in general appearance, but in the oxygen furnace a lance blows oxygen down from the top to produce high-quality steel at great speed.

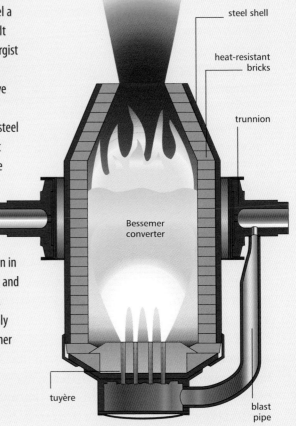

Steel gradually began to replace wrought iron and cast iron for making railroad tracks, large bridges, and steel-framed skyscrapers. Steelmaking became big business in the United States. But the decline of wrought iron was not immediate; as late as 1889, the Eiffel Tower in Paris was built from wrought iron.

## Steel alloys

In 1889, it was shown that steel alloyed with nickel made a very tough metal. It was adopted for armor-plating warships. In the 20th century, such alloy steels became very important. In 1912, the British metallurgist Harold Brearley (1871–1948) discovered stainless steels, which include chromium as well as certain types of nickel. Today, many kinds of alloy steels are produced for special purposes—for example, steel is alloyed with tungsten to give cutting tools very hard edges.

> **Alloy**
>
> Any material made of a blend of a metal with another substance to give it special qualities, such as resistance to corrosion or greater hardness.

## Other metals

Other existing metals and alloys also continued to play an important role during and after the Industrial Revolution. Brass was used increasingly from the 18th century onward for steam-engine cylinders, boilers, and domestic plumbing. The need for pure copper increased dramatically from about 1880 because of its use in electric wiring. Tin was in great demand from the 1850s, when canning food using tin-plated iron became a major industry. Lead became important in the 20th century for automobile batteries.

> **Curriculum Context**
>
> The curriculum requires that students can describe the properties of elements.

Over the past 100 years, other metals have come to the fore. Among the most important is aluminum. This only came into widespread use in the early years of the 20th century, but became indispensable in the aircraft, automobile, and electrical industries. Titanium, which is also extremely light but has better heat resistance, is used in many military aircraft. Another important new metal is tungsten, which has an extremely high melting point, and is used in lightbulbs.

# The Great Exhibition

**Staged in London in 1851, the first Great Exhibition was a world fair of exhibits designed to show the extent of human achievement. In practice, however, the emphasis was on science and technology, with more than half of the exhibits coming from Britain or the British Empire.**

Queen Victoria's husband, Prince Albert (1819–61), became president of the Royal Society of Arts in 1843. Six years later, he came up with the idea of mounting

an exhibition to show off the "Industries of All Nations." At that time—the end of the Industrial Revolution—manufacturing industry was still concentrated mainly in Britain, which was justifiably described as the workshop of the world. Queen Victoria (1819–1901) herself was the first person to put up money for the project. Immediately afterward, leading manufacturers also contributed funds to make up the £80,000 needed for the new building.

English architect and designer Joseph Paxton (1801–65) received the commission to design the great building

> **Curriculum Context**
>
> Students should know that in the past two centuries Europe has contributed significantly to the industrialization of Western and non-Western cultures.

Queen Victoria opens the Great Exhibition on May 1, 1851. In this contemporary print, Britannia and the British lion sit at the top of the frame, surrounded by symbols of the British Empire.

The Great Exhibition 17

to house the exhibition, which was put up in Hyde Park, London. (None of the 245 designs submitted in an earlier competition was considered suitable.) To demonstrate the technology of the era, Paxton designed a structure built entirely from prefabricated iron sections and glass, soon to be known as the Crystal Palace (a name thought up by the satirical magazine *Punch*). Paxton, the son of a farmer, had begun his working life as a gardener and went on to design hothouses and conservatories.

### Influential design
The Crystal Palace was "great" by any standards. It was 1,847 feet (563 m) long, 407 feet (124 m) wide, and over 100 feet (30.5 m) high. More than 3,000 iron columns and 2,000 girders supported 903,840 square feet (84,000 sq m) of glass—nearly enough to cover 17 football fields. The Crystal Palace influenced the design of many mainline railroad stations in Europe.

### The large and the small on show
There were about 14,000 exhibitors, showing more than 100,000 examples of manufactured goods. These included everything from machines as large as printing presses, railroad locomotives, and hydraulic machinery to items as small as cutlery and jewelry. The 560 exhibits from the United States included Cyrus McCormick's (1809–84) reaper and Samuel Colt's (1814–62) repeating revolver. Chewing tobacco was also included as something typically American. The French provided 1,700 exhibits. The exhibition remained open for 23 weeks, during which time it received more than 6 million visitors (most of them getting there by train). It also made a profit.

### Other international exhibitions
The Great Exhibition of 1851 was the first of several 19th-century international exhibitions (or expositions or world fairs, as some were called). The others were

> **Curriculum Context**
>
> The curriculum requires that students describe the factors that affect the purchase and use of manufacturing technology products and services.

## New York Crystal Palace

Modeled on London's Crystal Palace of 1851, but on a smaller scale, the New York Crystal Palace was constructed for the New York World's Fair in 1853. Like Joseph Paxton's building, it had modular prefabricated units made from cast iron with lots of glass to let in plenty of light. Wealthy visitors attended the exhibition in their own carriages; but because there was no railroad nearby, horse-drawn omnibuses and crowded horse-drawn trams (which can be seen in the foreground) carried most of the thousands of people to and from the site.

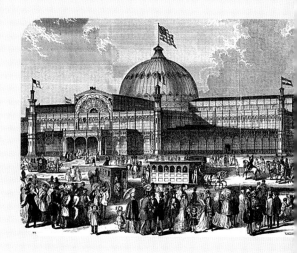

held elsewhere in Europe and North America—for example, in Vienna, Paris, New York, and Chicago. One of the grandest of the follow-up exhibitions was held in Chicago between May 1 and October 30, 1893. The 400th anniversary of the discovery of America was reflected in its title: the World's Columbian Exhibition. A team of East Coast architects had planned the layout three years earlier, resulting in a collection of 150 buildings that became known as the White City. With smooth white façades constructed in the classical style, they were built as the so-called Court of Honor around an artificial lagoon that opened off Lake Michigan. A double row of columns led down to the lake. The unified style of the exhibition's buildings began the Beaux-Arts period of American architecture that dominated major city centers for the next 40 years.

### Architecture

The science and art of designing and erecting buildings.

The original Crystal Palace was taken down in 1852 and rebuilt on a hill at Sydenham, south of London, where it remained in use for exhibitions and other shows until it was destroyed by fire in 1936. The London district where it once stood is still known as Crystal Palace.

# Genetics and Mendel

Today, genetics is one of the major scientific disciplines. It has applications in agriculture, biology, medicine, and even law enforcement. But it had very humble beginnings in the garden of a remote monastery in Austria, where the monk Gregor Mendel experimented by growing pea plants.

Gregor Mendel (1822–84) was born at Heinzendorf in Austrian Silesia (now Hyncice in the Czech Republic). He studied in college before entering the Augustinian order in 1843. By 1868, he was abbot of the monastery at Brünn (now Brno). He became interested in hybrids and began to breed pea plants in 1856. In the next six years he grew 30,000 plants, which he fertilized artificially by transferring pollen from the flowers of one plant to those of another. For example, he crossed tall plants with short plants. He then counted the number of tall and short plants in the next and later generations. He found that all first-generation plants were tall, but that the second generation contained tall plants and short plants in the ratio of 3 to 1.

**Curriculum Context**

Students should appreciate that much can be learned about the internal workings of science and the nature of science from the study of individual scientists.

Mendel concluded that every plant receives two "factors" of inheritance, one from each parent. In the first generation of peas in the above example, each plant receives one factor for tallness from the tall parent and one factor for shortness from the short parent. But all the offspring are tall, since the tallness factor is dominant over the shortness factor (which is described as recessive). Recessive factors can, however, make themselves apparent when two occur in a single individual, as in short plants of the second generation.

**Curriculum Context**

The curriculum requires an understanding of the basic laws of genetics and inheritance.

## Mendel's laws

These observations led Mendel to propose two laws. The law of segregation states that the two factors controlling each hereditary characteristic segregate

The Ages of Steam and Electricity

The diagram shows how Mendel's laws predict the colors produced by crossing purple-flowered peas with white-flowered peas. Purple is dominant, and the first generation (known as F1 phenotypes) produced by planting seeds from the cross are all purple. But when this generation is interbred, the second generation (F2 phenotypes) has purple or white flowers in the ratio 3 to 1.

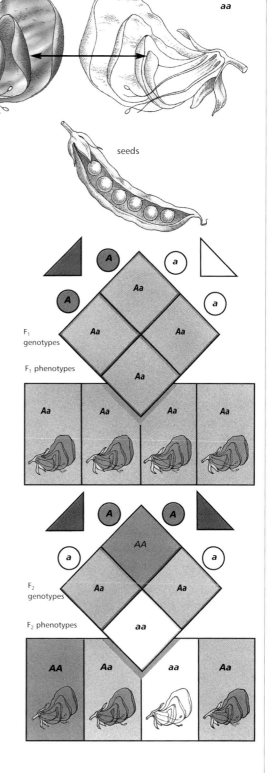

and pass into separate germ cells (egg and sperm). The law of independent assortment states that the pairs of factors segregate independently of each other during the formation of germ cells. In 1865, Mendel reported his results to the Brünn Natural History Society, and he published them a year later in the society's journal. Nobody took much notice, but Mendel's interest in botany continued, although with the increase in monastical duties, science ceased to occupy the central position in his life.

Mendel's "factors" are now called alleles, which are alternative forms of a gene. In any one body cell of an organism, there are two alleles of each gene, one inherited from each parent, which occupy the same place on a chromosome. Usually one allele is dominant, and the other is recessive. A germ cell (gamete)—egg or sperm—has only one allele. When egg and sperm combine at fertilization, the two alleles come together in a new individual that inherits characteristics from each parent. The appearance of the new individual, however, depends on which characteristic (if any) is dominant.

## Confirming Mendel's findings

In the 1890s, after Mendel's death, several European biologists independently studied inheritance in plants. In the Netherlands, Dutch botanist Hugo de Vries (1848–1935) came up with results identical to Mendel's, and his accidental discovery of Mendel's obscure publication prompted him to announce his results in 1900. De Vries' publication in turn led German botanist Karl Correns (1864–1933) and Austrian botanist Erich von Tschermak-Seysenegg (1871–1962) to publish their observations, which also confirmed that Mendel had been right all those years before. Between them, these four scientists founded the science of genetics.

## Principles of heredity

The diagram on page 21 shows the principles of heredity using pea plants (as Mendel did). We start by crossing a purple-flowered pea with a white-flowered pea. The seeds resulting from this cross are sown to produce first-generation hybrids, which have what biologists call the F1 phenotype. (The phenotype is the appearance of a plant or animal that results from its genetic makeup, known as its genotype.) All of these peas have purple flowers. This can be explained if we assume that the allele for purple (A) is dominant over the allele for white (a). If we allow the F1 generation to self-fertilize, and we plant their seeds, we get flowers of both colors. But purple outnumbers white in the ratio of 3 to 1. This is because among the F2 phenotypes one-quarter are AA and purple in color, two-quarters (one-half) are Aa and also purple (since A is dominant), and one-quarter are white (aa). White turns up only when two a recessive alleles occur together.

> **Curriculum Context**
>
> Students should understand that scientists' findings must be clearly reported to enhance opportunities for further investigation.

> **Curriculum Context**
>
> Students are expected to distinguish between dominant and recessive traits.

# The Calico Cat

All calico cats are females. The diagram shows how the mixture of black and ginger colors comes about. The genes for black and ginger pigments are carried on the X chromosome. Black (B) is dominant over ginger (b). Females have two X chromosomes, so they can be black (BB), ginger (bb), or calico (Bb). Males have only one X (and one Y), so they can only be black (B) or ginger (b). Males can never be calico—which requires one black (B) and one ginger (b) allele—because they only have one X chromosome. So when a calico female has kittens with a ginger father, three color types are possible: ginger males or females, black males, and calico females.

Sometimes the gene for a characteristic is located on one of the sex chromosomes (in mammals, these are named X and Y). For example, a gene on the X chromosome prevents color blindness when it is functioning properly. If one allele is defective, however, the condition can develop. More males than females have this disorder, since they have only one X chromosome. In females, however, both alleles—one on each X chromosome—would need to malfunction before they developed the condition. The same applies with the blood clotting disorder known as hemophilia.

> **Gene**
>
> The basic unit of inheritance that controls a characteristic of an organism.

# The electric telegraph

The telegraph was the first technological means of instant communication over distances beyond the range of the human voice. Before then, visual signals, such as the smoke signals made by some Native American peoples, were used, while institutions such as the British navy employed flags or pivoted semaphore arms, rather like railroad signals.

**Curriculum Context**

The curriculum requires that students should develop an understanding of electricity.

Darkness and poor visibility make a visual signal useless. But when it works, it is the fastest form of communication—signals travel between sender and receiver at the speed of light. Almost as fast is an electric current traveling along a wire. In 1804, shortly after the invention of the electric battery by Italian physicist Alessandro Volta (1745–1827), Catalan scientist Don Francisco Salvá i Campillo (1751–1828) devised a system that used 25 wires—one for each letter of the alphabet (no K). Each wire was connected to an electrode immersed in a tube of acidified water. A single wire interconnected the other electrodes in the tubes, and went back to the sender. When the sender connected this wire and one of the other wires to a battery, the current at the receiver caused electrolysis of the water. Bubbles then appeared at the electrode, identifying the chosen wire that represented a letter of the alphabet.

In 1809, German physicist Samuel von Sömmering (1755–1830) made a similar electrolytic telegraph, which operated over a distance of 1.9 miles (3 km) and required 35 wires. Then in England, in 1816, inventor Francis Ronalds (1788–1873) modified the system so that it needed only two wires to operate, and he offered his invention to the British Royal Navy. The navy was unimpressed with the invention, however, preferring to keep their antiquated mechanical semaphore method instead.

**Semaphore**

A method or device used for the visible transmission of messages, using lights, flags, or pivoted arms.

The Swiss physicist George Louis Le Sage experimenting with one of the first electric telegraphs in 1774. Its range was limited to the distance between two rooms in his house.

## Improving the system

Discoveries in physics provided further advances in the development of the electric telegraph. In 1820, Danish physicist Hans Ørsted (1777–1851) found that an electric current flowing along a wire deflected a pivoted magnetic needle placed nearby. By 1835, Joseph Henry (1797–1878) had made an experimental telegraph using pulses of electricity to represent coded letters. The pulses caused a piece of iron to "click" as it responded to an electromagnet at the receiver. Samuel Morse (1791–1872) later developed the idea.

Meanwhile, in 1832 Russian inventor Pavel Schilling (1786–1837) used Ørsted's discovery to make the first magnetized needle telegraph. It used six wires, and the electric current magnetized coils that deflected needles mounted above them. Schilling's invention was hardly noticed outside St. Petersburg, but German physicists Karl Gauss (1777–1855) and Wilhelm Weber (1804–91) saw his demonstration. In 1833, they sent signals over 1.9 miles (3 km), using a form of two-wire,

> **Curriculum Context**
>
> The curriculum requires that students should be aware that new ideas and inventions often affect other people.

A novelty trading card depicting a five-needle telegraph.

single-needle telegraph. In England, four years later, physicists William Cooke (1806–79) and Charles Wheatstone (1802–75) patented a needle telegraph. It used five needles that indicated letters on a diamond-shaped board in various combinations. It required six wires (five for the needles and one for the return current). This telegraph was installed on a section of the Great Western Railway in 1838. In 1843, they reduced the number of wires to three, and by 1845 they had simplified the receiver so that only one needle was required. By 1852, nearly 4,040 miles (6,500 km) of Britain's railroads were equipped with telegraphic communications.

## Messages in code

Morse demonstrated his single-wire telegraph in 1838, and its first commercial application was in 1844 over the railroad line along the 37 miles (60 km) between the cities of Washington and Baltimore. In fact, his telegraph was merely an improvement of Henry's ideas, but Morse's great contribution to communication was to invent the code of dots and dashes that became universally accepted for sending telegraph messages (and, later, radio messages as well). The final version of the Morse code was mainly the work of Morse's assistant Alfred Vail (1807–59).

## Linking the world

Soon, telegraph lines criss-crossed the countryside in North America as well as in Europe, and cables were laid beneath areas of water—in 1845 across New York Harbor, and in 1851 across the English Channel. In 1855, English-born American inventor David Hughes

A 19th-century artist's interpretation of the 20th century, highlighting innovations such as the telegraph.

(1831–1900) produced a printing telegraph. The sender tapped the message on a keyboard, and a similar machine at the receiver automatically typed the message as it arrived, letter by letter. In the United States in 1856, The New York and Mississippi Valley Printing Telegraph Company changed its name to Western Union Telegraph Company to signify the joining of telegraph lines from west to east. From that time onward the telegraph became the main method of local and international communications until it was superseded by telephone and radio.

> **Curriculum Context**
>
> Students are expected to describe the basic processes used in communication.

# Telephones

**The telephone was to become a momentous instrument in the history of communication. Not only could people now hear each other's voices clearly across great distances, but the process was almost instantaneous. The telephone truly linked the world.**

> **Curriculum Context**
>
> Students should be able to identify the characteristics of sound.

The first telephone was demonstrated in 1861 by the German inventor Philipp Reis (1834–74), but he used his invention largely as a teaching aid to illustrate how sound travels in waves. Alexander Graham Bell (1847–1922) and Elisha Gray (1835–1901) both filed patents for devices for sending speech down wires on the same day—February 14, 1876. By October of that year, Bell, a Scotsman who had emigrated to the United States, had demonstrated his device over a distance of 2 miles (5 km). Bell offered his patent to the Western Union Telegraph Company, but they saw no future in the telephone and turned him down. At first, Bell was almost the only person to realize the commercial possibilities of the telephone, and he set up his own telephone company. By 1887, however, there were more than 150,000 telephones in the United States.

## Louder signals

Speech quality with Bell's original telephone was good, but calls over more than a short distance were faint because the microphone in the telephone did not amplify (strengthen) the signal. Then, in 1886, the carbon microphone was produced by American inventor Thomas Edison (1847–1931). This microphone has small granules of carbon in the mouthpiece that vibrate

Operators working at a central telephone exchange in Paris toward the end of the 19th century.

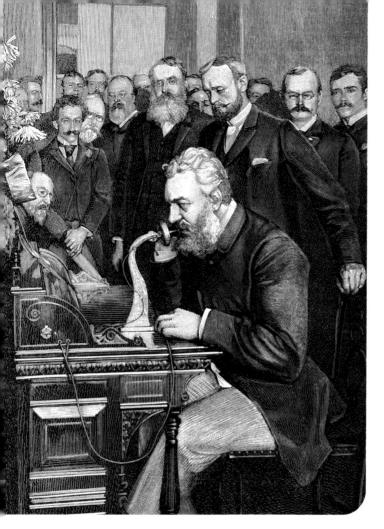

Alexander Graham Bell (seated) makes a call during the inauguration of a long-distance telephone line.

in time with the speech pattern, converting the sound waves into electrical energy that can be transmitted down the wire. Telephones with carbon microphones installed could now make long-distance calls, though at the expense of slightly poorer sound quality.

## Connecting cities by telephone

Another important development in 1886 was the creation of the American Telephone and Telegraph Company (AT&T), a subsidiary of Bell's own company. From 1887, AT&T opened trunk (long-distance) telephone lines across the United States, and by the start of the 20th century most towns in the East had been connected.

### Curriculum Context

Students should be aware that occasionally there are advances in technology that have important and long-lasting effects on society.

# Submarines

The first recorded attempt at building a submarine dates from about 1620. A Dutchman named Cornelis Drebbel (1572–1633) covered a rowing boat with greased leather. It had waterproofed holes covered with leather flaps for the oars. He demonstrated the vessel—the first of three—to his patron, King James I of England, maneuvering it below the surface of the Thames River in London.

After Drebbel's experiment, the next well-documented attempt to produce a submarine took place in North America. In 1776, student David Bushnell (c.1742–1824) built a barrel-shaped, one-man submarine called *Turtle*. It had a rudder, two hand-operated propellers (one for up-and-down movement and one for forward movement), and a hand pump for pumping out the water ballast to allow the craft to surface. Attached to the outside of the "barrel" was a container of gunpowder that could be stuck to the hull of an enemy ship. It had hooks that could be released by pulling cords from inside. Bushnell's submarine was put to the test during the American Revolution, when it attacked a British ship in New York Harbor. The attack failed.

### Successes and more failures

A slightly more successful submarine was built by American engineer Robert Fulton (1765–1815) in France in 1801. Called *Nautilus*, it was a slow, hand-cranked machine built of copper plates on an iron framework. At 21 feet (6.4 m) long, it could carry four people, and it could remain underwater for three hours before the oxygen ran out. It successfully sank an "enemy" ship during one demonstration.

In 1851, German soldier Wilhelm Bauer (1822–75) built an unsuccessful craft called *Brandtaucher* (Fire Diver). It had a crew of three, two of whom worked a treadmill to provide propulsion. His larger *Diable-Marin* (Sea

**Curriculum Context**

Students should know that the process of respiration requires the presence of oxygen.

*The Ages of Steam and Electricity*

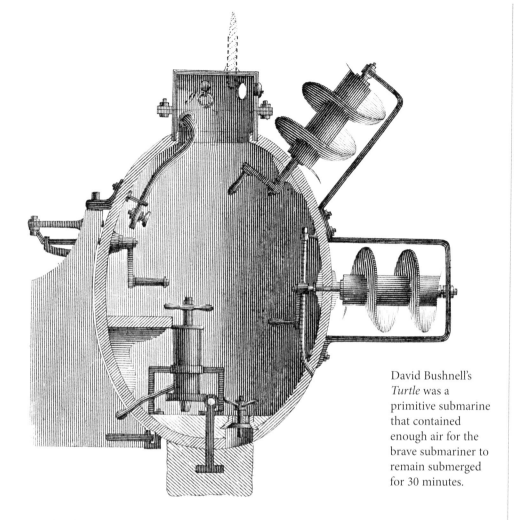

David Bushnell's *Turtle* was a primitive submarine that contained enough air for the brave submariner to remain submerged for 30 minutes.

Devil) of 1855—a 52.5-foot (16-m) submarine with a crew of 16 men—was more successful. It made over 130 dives before it sank.

In the United States in 1863, American engineer Horace Hunley (1823–63) copied Robert Fulton's design, and built a submarine for the Confederacy. The submarine was armed with a ram carrying an explosive charge, and eight men had to work a long crankshaft to turn the propeller. Unfortunately, Hunley was killed along with the rest of the crew when the vessel sank during its second trial. However, the submarine was raised, and successfully attacked the Union ship *Housatonic* in Charleston Harbor in 1864, only to sink again because its ram remained stuck in the ship's hull.

The submarine *Nordenfeldt 1* took up to 12 hours to generate enough steam for submerged operations to begin, and took about 30 minutes to dive.

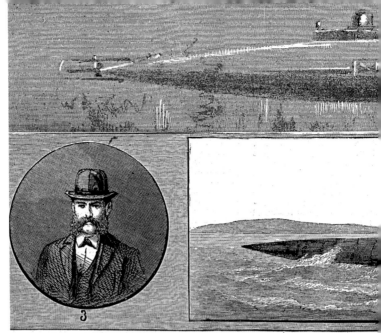

### Compressed air

Air under greater pressure than the air in the general environment, especially when used to power a mechanical device, or to provide a portable supply of oxygen.

### New forms of propulsion

In 1863, the French engineer Simon Bourgeois built an experimental submarine, called *Le Plongeur* (The Diver), with an engine that was powered by compressed air. However, the first real French success came in 1888, when engineer Gustave Zédé (1825–91) completed a vessel called *Le Gymnote* for the French navy. It was 55.8 feet (17 m) long, and was powered by a 51-horsepower electric motor that drove a 5-foot (1.5-m) propeller. It achieved a speed of almost 7 miles per

hour (11 km/h) on the surface and nearly 5 miles per hour (8 km/h) when submerged. Other inventors experimented with steam propulsion. The boiler fires in the wooden-hulled *Resurgam*, built by English clergyman George Garrett (1852–1902), had to be extinguished before a dive. Underwater, the vessel relied entirely on stored steam. The 66-ton (60-tonne) *Nordenfeldt I*, built in 1882 by the Swedish gunsmith Thorsten Nordenfeldt to Garrett's design, used a similar system and also carried a torpedo tube.

**Curriculum Context**

The curriculum requires that students can describe how and why technology evolves.

### Gasoline fuel

A volatile, flammable liquid composed of distillates obtained in the petroleum refining process with a boiling range of 30°C to 200°C (85°F to 390°F), and ignited in engines by a spark.

### Diesel fuel

A fuel composed of distillates obtained in the petroleum refining operation; it has an ignition temperature of 540°C (1,000°F), and is ignited in engines by the heat generated from compressed air.

### Toward the modern submarine

Irish–American teacher John Holland (1840–1914) provided the solution to the propulsion problem. Financed by revolutionaries who wanted to achieve Irish independence from the British, Holland built submarines that used gasoline engines for surface propulsion and electric motors when submerged. They included the 21-ton (19-tonne) submarine *Fenian Ram*, with a three-man crew, launched in 1883.

A series of developments resulted in the launch, in 1898, of *Holland VI*, a 52.5-foot (16-m) vessel that could travel at 7 miles per hour (11 km/h) under water. It carried a self-propelled torpedo and a deck gun. The U.S. Navy bought the vessel in 1900 and renamed it USS *Holland*. Later "*Hollands*" were fitted with a periscope, designed in 1902 by American engineer Simon Lake (1866–1945), who in 1897 built the seagoing submarine *Argonaut*. Holland sold six boats to the U. S. Navy and received orders from the British, Japanese, and Russian navies.

The launch of a diesel-engined submarine in England, in 1908, was another major turning point in the submarine's evolution. Diesel remained the standard means of power until 1955, when U.S. Admiral Hyman Rickover (1900–86) built the first nuclear-powered submarine—*Nautilus*. The use of nuclear power overcame the need to surface and recharge batteries, and *Nautilus* became the first submarine to sail submerged around the world.

### Underwater weaponry

Although U.S. inventor David Bushnell devised the first mine, an underwater explosive device, in 1775, it was not until 1843 that his compatriot Samuel Colt (1814–62) detonated one under water with the help of an electric charge, and transformed it into a potent weapon. The torpedo, meanwhile, was the joint invention of the Englishman Robert Whitehead (1823–1905) and the

# How a Submarine Works

A submarine has a twin hull. The space between the two hulls holds water that acts as ballast to keep the vessel submerged. When the submarine floats at the surface, the ballast tanks are empty. To make the submarine sink, the lower valves are opened, and seawater enters the ballast tanks (at the same time, the upper valves open to let the air out of the ballast tanks). As a result, the submarine becomes heavier than the water it displaces, and it sinks. In order to make the submarine rise again, the upper valves are closed, and compressed air forces water out of the ballast tanks through the lower valves. The submarine can be "trimmed," with the right amount of ballast in the tanks to keep it floating just below the surface of the water.

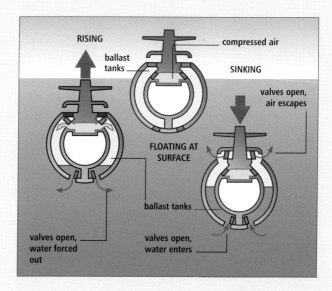

Austrian Giovanni Luppis (1813–75). In 1866, Whitehead built an underwater missile that was 14 feet (4 m) long, and driven by compressed air at a speed of 6 knots (7.2 miles per hour or 11 km/h) for 700 yards (640 m). Although widely used in World War I (1914–18), technical problems made torpedoes unreliable weapons. These were overcome in the 1930s by the Japanese, who invented the Long Lance torpedo, with a speed of 49 knots (56 miles per hour or 91 km/h) and a range of 11 miles (18 km). The sophistication of these weapons has increased with the introduction of electronic guidance systems.

The addition of nuclear ballistic missiles in the 1960s, meanwhile, created submarines capable of destroying cities thousands of miles away.

### Curriculum Context

The curriculum requires that students can describe interrelationships between temperature, particle number, pressure, and volume of gases contained within a closed system.

# The Periodic Table

**By 1869, following a spate of discoveries using the new techniques of electrolysis and spectroscopy, there were 63 known chemical elements. They were the building blocks from which, in that year, Dmitri Mendeleev constructed his famous Periodic Table.**

Dmitri Mendeleev (1834–1907) was born in Tobolsk, Siberia. His father died when Mendeleev was 13, but his mother was nevertheless determined that Dmitri should receive good schooling. He won a place at the Pedagogical Institute in St. Petersburg, and qualified as a teacher in 1855. He later studied chemistry at the University of St. Petersburg and at the University of Heidelberg in Germany. He took up a university post at St. Petersburg, and in 1869 began writing a textbook on chemistry (in those days, inorganic chemistry).

### Organizing the elements

Wishing to find some order in the apparent jumble of chemical elements, he wrote the name of each on a card and then tried to deal sets of "hands," like dealing out a deck of playing cards. He arranged the elements in order of increasing atomic weight (the average mass of each atom of the element). He found that if he started a new row of cards every eighth element, those with similar chemical properties fell one above the other in columns.

Looking at the rows, he saw that properties tended to recur along each row—there was what he called a "periodicity" in the properties. He named his new grid of rows and columns the Periodic Table. He also had the foresight to include in the table additional "missing" elements that were still to be discovered. He even predicted the chemical and physical properties of these elements—for example, their atomic weights and melting points.

> **Curriculum Context**
>
> The student should understand that types and levels of organization, such as the Periodic Table, provide useful ways of thinking about the world.

The Russian chemist Dmitri Mendeleev in his laboratory. As well as his work on known elements, he was able to predict the existence of others yet to be discovered.

In 1875, "eka-aluminum" (in a space below aluminum) was discovered by French chemist Paul Lecoq de Boisbaudran (1838–1912) and named gallium; in 1879, Swedish chemist Lars Nilson (1840–99) discovered "eka-boron" (below boron), which was named scandium; and in 1886, "eka-silicon" (below silicon) was discovered by German chemist Clemens Winkler (1838–1904) and given the name germanium. Mendeleev's predictions were being fulfilled. By 1914, there were only seven remaining gaps in the Periodic Table up to element 92.

### Atomic structure

The atomic number is the total number of protons in an atom of any element, and the modern Periodic Table is better described as being arranged in order of atomic number. In recent times, chemists have introduced the term "neutron number" (number of neutrons in the atom's nucleus), and they now refer to

> **Curriculum Context**
>
> Students are required to summarize the historical development of the Periodic Table in order to understand the concept of periodicity.

> **Curriculum Context**
>
> Students should be able to relate the chemical behavior of an element to its placement on the Periodic Table.

atomic weights as relative atomic masses. Mendeleev could not explain the reason for the periodicity of the elements. That had to await an understanding of the structure of atoms, particularly the way in which electrons arrange themselves around the nucleus of an atom. By the second quarter of the 20th century, chemists realized that the Periodic Table reflects the atomic structures of the elements as electrons fill up shells surrounding the nucleus. All chemical reactions involve electrons, particularly an element's outer electrons. The Periodic Table enables chemists to predict more accurately what reactions are possible, which are likely to take place under ordinary laboratory conditions, and which will require extra effort such as higher temperatures, higher pressures, or catalysts to make them take place. Mendeleev received the

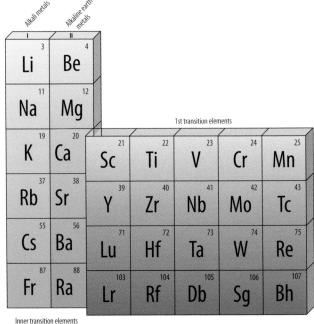

The Periodic Table. The system of using a letter or letters to denote the symbol for each element (for example, Fe = iron) was introduced by Swedish chemist Jöns Berzelius (1779–1848) in 1818.

ultimate scientific accolade in 1955 when the newly discovered element 101 was named mendelevium in his honor.

Today's Periodic Table contains about 50 more elements than in Mendeleev's time. They are arranged, in order of their atomic numbers, in seven horizontal "periods" of varying length. Two very long series, of 14 elements each—the lanthanides following lanthanum (57) and the actinides following actinium (89)—are shown in separate lines below. The most recently discovered elements, such as seaborgium (Sg) and bohrium (Bh), are named for famous 20th-century scientists. Others are named for places: berkelium (Bk) for the University of California, and dubnium (Db) for the Russian Joint Institute for Nuclear Research at Dubna, near Moscow.

> **Curriculum Context**
>
> The curriculum requires that students can describe the physical and chemical characteristics of an element using the Periodic Table.

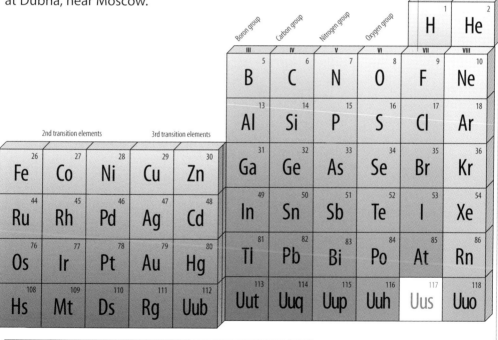

# Mapping the Moon and Mars

Astronomers have drawn maps of the Moon ever since Galileo pointed his first telescope at the heavenly body in 1610. Then, with the development of more powerful telescopes, Mars began to grab the attention of astronomers. Most noted of them was Giovanni Schiaparelli, whose apparent reference to canals on Mars was the start of much debate among astronomers.

The most powerful telescope used by Galileo Galilei (1564–1642) gave him an image of the Moon that was at best six times the size it appears to the naked eye. Even so, he made some detailed drawings, which

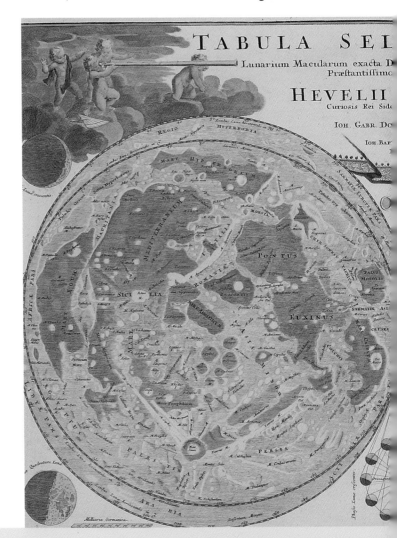

proved that the rather mottled surface features are, in fact, caused by mountains and craters. The Flemish cartographer Michael Langrenus (1600–75) published the first detailed map of the Moon in 1645. He introduced the idea of naming lunar mountains and other features for astronomers and famous scientists. For example, he named one of the prominent craters Hipparchus, for the Greek astronomer Hipparchus of Nicaea (c.190–125 B.C.). Like other people of his time, he thought that the dark areas were expanses of water, and called many of them *mare* (Latin for sea). However, astronomers still argued about the origin of the craters—were they ancient volcanoes, or were they caused by meteorites smashing onto the surface?

### Curriculum Context

Students should be able to identify components of the Solar System including the Sun, planets, and moons.

The Moon is close enough to Earth for some detail to be visible with the naked eye. But it was the invention of the telescope that made maps like this one a possibility.

Mapping the Moon and Mars

By 1836, English astronomer Francis Baily (1774–1844) confirmed the existence of the large mountains by drawing and analyzing the phenomenon that came to be known as Baily's beads. Observing the Moon during an eclipse of the Sun, he noticed that just as the Moon appears to cover the Sun's disk, a "row of lucid points, like a string of bright beads" forms around the curved edge of the Moon. Baily correctly interpreted them as being caused by the Sun shining through valleys between the tall mountains at the Moon's rim.

### Lunar photographs

In 1839, pioneer French photographer Louis Daguerre (1787–1851) included the Moon in a daguerreotype, and in 1840, English-born American scientist John Draper (1811–82) set out deliberately to make daguerreotypes of the Moon. Better and faster-acting photographic emulsions made Moon photography easier, but hand-drawn maps based on observation continued to be produced until the late 19th century, including one drawn by German astronomer Wilhelm Lohrmann (1796–1840), and published in 1878 by Johann Schmidt (1825–84). Closeup Moon photography only came in the 20th century. In 1945, members of the U.S. Signal Corps bounced radar signals off the Moon, and detailed pictures were taken in the 1950s and 1960s by the Soviet *Lunik* and *Luna* probes and NASA's *Ranger* series missions.

### Observing Mars

Mars has always held a fascination for people on Earth, particularly the possibility that there might be life on the "Red Planet." In 1666, Italian astronomer Giovanni Cassini (1625–1712) first pointed out the existence of "ice caps" at the north and south poles of Mars—a fact confirmed in 1704 by Italian cartographer and astronomer Giacomo Maraldi (1665–1729), who mapped the way in which they vary in size over the Martian seasons. (Cassini also drew a map of the Moon

> **Daguerreotype**
>
> The earliest photographic process, in which a copper plate is coated with silver and left exposed to iodine vapor, creating light-sensitive silver iodide. After use in a camera, the plate is "developed" in mercury vapor and "fixed" in salt, producing a mirror image of a scene.

> **Curriculum Context**
>
> The curriculum requires that students should appreciate that new techniques and tools can provide new evidence to guide scientific inquiry.

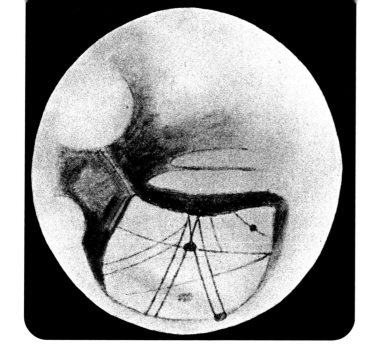

One of Percival Lowell's images of Mars, depicting what he claimed were water "canals," and supporting his belief that intelligent life existed on the planet.

based on eight years of observations, which acted as a standard reference for more than a century.) Dark areas on the planet's surface were first thought to be seas, and then dried-up seabeds.

### Life on Mars?

Then, in 1877, came the big red herring concerning the Red Planet. Italian astronomer Giovanni Schiaparelli (1835–1910) drew a map of Mars on which he marked dark lines representing what he called "channels." Unfortunately, the Italian word *canali* was translated into English as "canals." Their existence led American amateur astronomer Percival Lowell (1855–1916) to believe they were an irrigation system built by Martians to carry water from the melting ice caps to the dry equatorial regions. Lowell took the first photographs of Mars in 1905, using the telescope at his observatory in Arizona. Modern astronomers regard the Martian canals as a historical curiosity, or at best an optical illusion, an opinion that was confirmed by the American *Mariner 4* probe of 1965 and the *Mars* series of 1971. The "ice" in the ice caps is now thought to consist mainly of frozen carbon dioxide.

> **Curriculum Context**
>
> Students should be aware that as scientific knowledge evolves, major disagreements are eventually resolved through interactions between scientists.

# Germs and disease

During the middle of the 19th century, scientists finally realized that germs cause most diseases, and people no longer blamed "evil spirits" or "bad air." With improved microscopes and new laboratory techniques, scientists set about tracking down these sometimes deadly microorganisms.

**Curriculum Context**

The curriculum requires an understanding of the role of bacteria and viruses in causing diseases such as in streptococcus infections and diphtheria.

As early as 1546, Italian physician Girolamo Fracastoro (c.1478–1553), in his book *De contagione et contagiosis morbis* (On Contagion and Contagious Diseases), suggested that germs are the cause of disease. Nobody took much notice, even after 1676, when Dutch scientist Antonie van Leeuwenhoek (1632–1723) first saw bacteria, using a homemade microscope.

Then, in 1840, German pathologist Jacob Henle (1809–85) put forward the idea that infection is caused by parasitic organisms—the so-called "germ theory" of disease, which was later proposed independently by French chemist Louis Pasteur (1822–95). In 1877, German bacteriologist Robert Koch (1843–1910) showed that bacteria could be stained to make them easier to study under a microscope. Seven years later, Danish physician Hans Gram (1853–1938) used this idea as a means of classifying bacteria, which since then have been dubbed either Gram-positive or Gram-negative depending on their capacity to absorb a special stain. Bacteriologists also classify bacteria according to their shapes: coccus (round), bacillus (oval), spirochete (spiral), and so on.

**Curriculum Context**

Students should be aware that some advances in science have long-lasting and profound effects on society.

## Making quick discoveries

Once biologists knew what bacteria looked like, the hunt was on, and scientists who handled cultures of infectious diseases often put themselves at risk. Results came quite quickly. In 1880, German bacteriologist Karl Eberth (1835–1926) found the bacillus causing typhoid.

Louis Pasteur used rabbits and other experimental animals in his studies of bacteria. He grew the bacteria in nutrient cultures, kept in a basement room equipped with a boiler and hot pipes to keep them warm.

Germs and disease

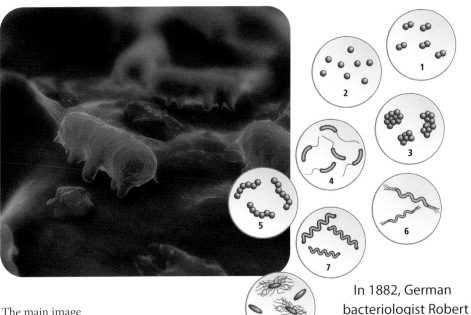

The main image shows *Salmonella* bacteria, which are among the most common causes of food poisoning in humans. These oblong bacteria are classified as a type of bacillus. The various common shapes (above right) are: monococcus (1), diplococcus (2), staphylococcus (3), vibrio (4), streptococcus (5), spirillum (6), spirochete (7), and bacillus (8).

### Curriculum Context

Students should be able to describe the structure and functions of viruses.

In 1882, German bacteriologist Robert Koch (1843–1910) found the bacterium that causes tuberculosis, and German bacteriologists Friedrich Löffler (1852–1915) and Wilhelm Schütz (1839–1920) identified the cause of the animal disease glanders. In 1897, Danish veterinarian Bernhard Bang (1848–1932) discovered a bacillus that causes abortion in cattle, and the Japanese bacteriologist Kiyoshi Shiga (1871–1957) found the cause of endemic dysentery.

## Harmful protozoans and fungi

Bacteria are not the only parasitic microorganisms to cause human diseases. Protozoans, for example, include the trypanosomes that cause sleeping sickness and Chagas' disease, the amebas that result in amebic dysentery, and the *Plasmodium* parasite responsible for malaria. Some microscopic fungi produce diseases that affect the skin or lungs. Most of these microorganisms were tracked down by 19th-century microbiologists.

## Fighting viruses

In 1897, Dutch microbiologist Martinus Beijerinck (1851–1931) showed that the microorganism causing

# Viruses

Viruses are too small to be trapped by a filter that retains bacteria. For this reason, they evaded discovery until the very end of the 19th century. Even so, nobody actually saw a virus until after the invention of the electron microscope in the late 1930s. These microorganisms turned out to have various shapes, and to consist of an outer "container" of protein holding a molecule of DNA (deoxyribonucleic acid) or RNA (ribonucleic acid). Various structures may project from the outer surface. Viruses cannot multiply outside a living cell; but once they force their way into a cell, they take it over, and make it rapidly reproduce more virus particles. The particles break out, and quickly invade other cells. Shown below are: an adenovirus (A) that causes acute respiratory disorders; a bacteriophage (B) that attacks bacteria; and HIV (human immune deficiency virus), which causes AIDS (C).

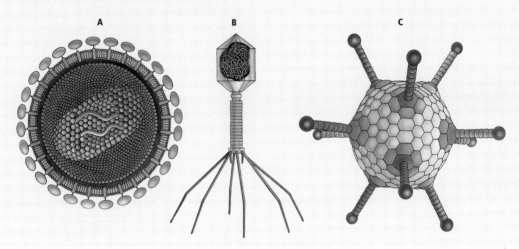

tobacco mosaic disease escapes through a filter that normally traps bacteria. He had discovered the first virus. A year later, the virus that causes foot-and-mouth disease in cattle was discovered. Since then, viruses have been found to be responsible for many diseases in humans, including influenza, polio, measles, and AIDS (acquired immune deficiency syndrome). Almost as fast as bacteriologists found the bacteria that cause diseases, they developed vaccines against them, so that people could be injected and gain immunity. Vaccines for virus diseases proved more difficult, but now exist for all the disorders named above except AIDS.

### Vaccine

A preparation containing viruses or other microorganisms, introduced into the body to stimulate the formation of antibodies and build up immunity against infectious disease.

# The internal combustion engine

In a steam engine, which was the main power source for the Industrial Revolution, combustion of the fuel in order to boil water and make steam takes place outside the engine mechanism itself. It is an example of an external combustion engine. However, an engine is much more efficient if the fuel burns inside the cylinder, as in the internal combustion engine.

**Coal gas**

Flammable gas obtained in the destructive distillation of soft (bituminous) coal, often as a byproduct in the preparation of coke.

The Belgian engineer Étienne Lenoir (1822–1900) made the first successful engine to burn fuel internally, in 1859. It ran on coal gas, which was mixed with air, and sucked into the cylinder by the movement of a piston. When the piston was halfway along the cylinder, an electric spark ignited the gas-and-air mixture, which exploded, and forced the piston to the end of its stroke. As the piston moved back, gas and air were sucked in on the other side of the piston, and the process was repeated. It produced about 1 horsepower at its speed of 200 rpm (revolutions per minute). It worked in two jerky stages, and also needed a heavy flywheel.

Then, in 1862, French engineer Alphonse Beau de Rochas (1815–93) patented an engine that worked in four stages, or strokes. He did not build an actual engine; and when his patent expired, the idea was taken up by a German engineer, Nikolaus Otto (1832–91). In 1876, Otto made his first horizontal four-stroke gas engine. It had a port, or hole, in the cylinder that opened for a flame to ignite the fuel-and-air mixture. It produced 3 horsepower. The four-stroke cycle is still the basis of today's modern engines.

**Curriculum Context**

Students should know that in looking at the history of many peoples, one finds that engineers of high achievement are considered to be among the most valued contributors to their culture.

### The first gasoline engine

Otto's engine still used coal gas as a fuel. In 1867, the Austrian engineer Siegfried Marcus (1831–98) invented a carburetor—a device that vaporized liquid gasoline, and mixed it with air. Two German engineers, Karl Benz

(1844–1929) and Gottlieb Daimler (1834–1900), who had both worked for Otto, independently made the first gasoline-burning internal combustion engines in 1885. Both engines were used to develop motor cars and motor bicycles. Daimler's engine ran at a speed of 900 rpm, and used a red-hot platinum tube to ignite the fuel and a carburetor invented by his partner, German engineer Wilhelm Maybach (1846–1929). Called a surface carburetor, it produced a gasoline-and-air mixture by forcing a current of air across the surface of the gasoline. The engine built by Benz was slower (only 250 rpm), and produced less than 1 horsepower, but the car he used it in had many modern features, including coil ignition powered by a battery, and a distributor (a rotating contact-breaker switch geared to the engine).

### Platinum

A silvery-white, malleable, ductile, metallic element often used as a chemical catalyst; platinum is highly resistant to corrosion and tarnish.

## The diesel engine

By the end of the 19th century, the scientific principles of engines—the applications of thermodynamics—were being analyzed. One consequence of this study was the prediction that if a suitable fuel–air mixture is

An early motor vehicle driven by an internal combustion engine attracts stares of astonishment from bystanders.

## The Two-stroke Engine

Scottish mechanical engineer Dugald Clerk (1854–1932) invented the two-stroke engine in 1878. In a two-stroke engine, the need for separate strokes to let the fuel in and the exhaust out is avoided. That is because the piston also acts as a valve, allowing the fuel-and-air mixture in and the exhaust gases out of the cylinder. On the upstroke, the piston compresses the fuel-and-air mixture, which is ignited by a spark from the spark plug. At the same time, the piston closes off the exhaust port. The exploding mixture expands, forcing the piston down—the downstroke. Exhaust gases escape through the exhaust port, which is now not covered by the piston, and as the piston rises again, more fuel and air is sucked into the engine through the inlet port. The crankshaft converts up-and-down piston motion into rotary motion.

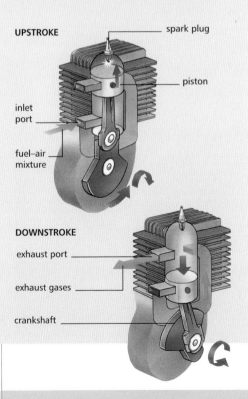

hot enough and compressed enough, it will combust spontaneously without the need for a spark. The first engineer to put the idea into practice was the Englishman Herbert Stuart (1864–1927), who in 1890 patented the first example of what is called a compression-ignition engine. Two years later, a German inventor patented a similar engine, and demonstrated it in 1897. His name was Rudolf Diesel (1858–1913), and since then this engine has been known as a diesel engine.

A diesel engine has advantages in many applications. It uses a less refined fuel than gasoline, and diesel fuel is a much denser and less flammable product than gasoline. Also, the engine needs no spark plugs or associated ignition system, and is 35 percent efficient in terms of fuel consumption, compared with 25 percent for the best gasoline engines.

### Rotary engines

The evolution of the internal combustion engine was still not complete. Like the steam engine before it, the early gasoline (and all diesel) engines were reciprocating engines. The oscillating pistons produced up-and-down (or side-to-side) movement, which had to be converted to a rotary motion for practically every application. There were, however, some "rotary" gasoline engines, developed very successfully as power plants for propeller-driven aircraft.

In 1929, German engineer Felix Wankel (1902–88) patented a revolutionary internal combustion engine—"revolutionary" in that it was truly rotary. The first prototype was made in 1956. A Wankel engine has a rotor (in section, it resembles a triangle with slightly curved sides) that rotates inside a "cylinder" shaped like a fat figure-eight. The geometry creates three separate areas that can be considered as combustion chambers. The engine has a four-stroke cycle using one or two spark plugs and two ports (one for fuel-and-air inlet and one for exhaust outlet).

> **Curriculum Context**
>
> Students should be able to describe how changes in technology affect engineering practices.

## The Four-stroke Engine

The gasoline engines in most cars use the four-stroke cycle. Unlike a two-stroke, a four-stroke engine has valves to open and close the ports that allow the fuel-and-air mixture into and the exhaust gases out of the cylinder. On the intake stroke, the inlet valve opens as the piston moves down. This sucks the mixture into the cylinder. On the compression stroke, the inlet valve closes and, as the piston rises, the fuel-and-air mixture is compressed. During the power stroke, the spark plug ignites the fuel, which explodes, and the expanding hot gases force the piston down. On the fourth stroke, the exhaust valve opens to allow exhaust gases to escape while the piston rises. The up-and-down motion of the piston is turned into rotary motion by the crankshaft.

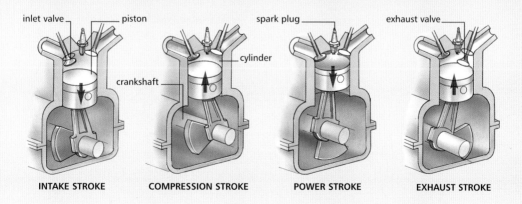

INTAKE STROKE  COMPRESSION STROKE  POWER STROKE  EXHAUST STROKE

# Sources of electricity

**Before 1800, scientists knew only about electricity that was stationary, or static. This static electricity consists of the positive or negative charge that an object possesses, usually produced by friction. Then an Italian nobleman produced an electric current—this was electricity on the move.**

**Curriculum Context**

Students should understand the concepts associated with electricity.

In 1791, Italian physician Luigi Galvani (1737–98) reported what he called "animal electricity." While dissecting a dead frog, he found that the animal's muscles twitched when he touched them with two different metals. Then, in 1800, Count Alessandro Volta (1745–1827), an Italian physicist, replaced the animal tissue with a disk of cardboard soaked in salt solution. He put a piece of copper or silver on one side, and a piece of zinc on the other side. Wires connected to the metal plates carried an electric current. Volta found he could obtain higher voltages by making a stack of such disks to form what became known as a voltaic pile, which was the first true battery.

## Improving the system

Today, scientists call such a battery a primary cell. The pieces of metal are called electrodes, and the solution between them is an electrolyte. In 1836, English chemist John Daniell (1790–1845) produced a more efficient primary cell. It had a zinc rod electrode dipped in dilute sulfuric acid, contained in an earthenware pot. He immersed the pot in a copper container (which acted as the other electrode) containing copper sulfate solution. In the Daniell cell, an electric current flows from the copper (the positive electrode, or anode) to the zinc (the negative electrode, or cathode). It gives a steadier current than Volta's cell and overcomes polarization—a build-up of hydrogen bubbles on the copper electrode that eventually stops the flow of electrons, and causes the voltaic pile to stop working.

Alessandro Volta took every opportunity to demonstrate his new battery, or voltaic pile. Here, he shows it to Napoleon I, the young emperor of France.

The Leclanché cell, a battery invented in 1866 by French engineer Georges Leclanché (1839–82), also avoids polarization. It, too, has a zinc cathode, but dips into an electrolyte of ammonium chloride solution. The anode is a carbon rod surrounded by manganese dioxide powder. Today's common type of dry battery has the same system, using a zinc case containing a paste of ammonium chloride, with a carbon rod surrounded by manganese dioxide down the center.

Sources of electricity 53

## Using other metals

The German chemist Robert Bunsen (1811–99) also made a zinc–carbon primary cell; using acid electrolytes, it produces 1.9 volts. The cadmium cell, invented in 1893 by English-born American electrical engineer Edward Weston (1850–1936), produces 1.0186 volts, and in 1908 the scientific community accepted it as a standard of voltage. Known as the Weston standard cell, it has mercury (cathode) and a cadmium–mercury mixture (anode) as its electrodes, with cadmium sulfate solution as its electrolyte. The Clark standard cell, invented by English electrical engineer Josiah Clark (1822–98) 21 years earlier, in 1872, uses zinc instead of cadmium.

> **Curriculum Context**
>
> The curriculum requires that students should be aware that technological changes often happen in small steps.

## Accumulators

A primary cell stops working once it is fully discharged. A different type of cell, called variously a secondary cell, a storage cell, or an accumulator, can be recharged. The lead–acid accumulator, devised in 1859 by a French chemist, Gaston Planté (1834–89), is the earliest and still the most commonly used kind. It has one electrode (or "plate") of lead and one of lead covered with lead oxide, which dip into a sulfuric acid electrolyte. This is the type of battery used in most automobiles. The alkaline nickel–iron accumulator, or Ni–Fe cell, is another type, invented in 1900 by American inventor Thomas Edison (1847–1931). When any accumulator becomes discharged, it can be connected to a supply of DC (direct current), and recharged. For example, while a car's motor is running, the battery is continually recharged.

> **Curriculum Context**
>
> Students should identify transformations occurring during energy production for human needs.

## Fuel cells

Both primary and secondary cells convert chemical energy into electrical energy. In doing so, they "consume" materials in the electrodes or electrolyte. A fuel cell, however, converts the chemical energy of a fuel directly into electrical energy. The first fuel cell,

# Primary and Secondary Cells

A primary cell, such as the dry battery for a flashlight (below left), can be used only until its chemicals run out, when it has to be thrown away (carefully). In this type of cell, the electrolyte is a paste of ammonium chloride and gum. The zinc cathode forms the case of the battery, and manganese dioxide and carbon surround the central carbon anode. A secondary cell, or accumulator (right), can be recharged when it runs down, and reused. It has lead and lead-oxide electrodes in a sulfuric acid electrolyte. When it is in use, sulfate ions react with the lead cathode to produce lead sulfate, and release electrons. At the anode, hydrogen ions from the acid and sulfate ions react with the lead oxide to produce lead sulfate and water. The reactions produce about 2 volts. To recharge the accumulator, current from an outside source is passed through the battery in the opposite direction. This has the effect of reversing the reactions at the electrodes, reforming lead and lead oxide. The accumulator is then ready for use again. Automobiles use batteries of this type.

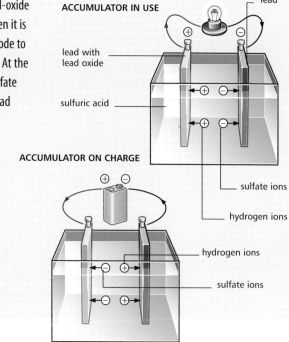

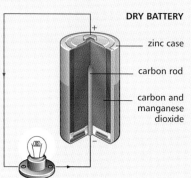

which "consumed" hydrogen and oxygen gases, was demonstrated in 1839 by Welsh physicist and judge William Grove (1811–96).

### Forever remembered

Of all these scientists, the best remembered is Alessandro Volta. He gave his name to the volt, which was adopted in 1905 by the International Electrical Congress and is now the SI unit of electric potential.

> **Curriculum Context**
>
> Students should be able to demonstrate the use of SI units.

# The invention of radio

**Radio is a method of communication that uses electromagnetic radiation known as radio waves, which travel at the speed of light. It used to be called "wireless" to distinguish it from telegraphs and telephones, which both needed wires running between the sender and the receiver to transmit their signals.**

### Curriculum Context

Students should understand that electromagnetic waves include radio waves (the longest wavelength), microwaves, infrared radiation (radiant heat), visible light, ultraviolet radiation, x-rays, and gamma rays.

Radio has its origins in the late 19th century. Danish physicist Hans Ørsted (1777–1851) had already established that electricity and magnetism were part of the same phenomenon (called electromagnetism), and in 1864 James Clerk Maxwell (1831–79), a professor of physics at Cambridge University in England, showed that energy could be transmitted in the form of an electromagnetic wave, not unlike the waves of the sea, but at the speed of light.

In 1888, the German physicist Heinrich Hertz (1857–94) produced electromagnetic waves (called Hertzian waves) using equipment that generated large electrical sparks that jumped between two metal spheres. Hertz and Maxwell had laid the foundations for "wireless telegraphy," or radio.

### The "coherer"

In 1890, French physicist Édouard Branly (1844–1940) devised the first way of detecting radio waves, using a device called a "coherer:" a sealed glass tube containing iron filings and an electrode at each end. When radio waves are present, the filings cohere (stick together), and conduct electricity sufficiently well to form part of a circuit. English physicist Oliver Lodge (1851–1940) improved the coherer in 1894, and used it together with a spark transmitter to send Morse code messages a distance of 490 feet (150 m). Russian physicist Aleksandr Popov (1859–1906) conducted similar experiments a year later.

### Morse code

A telegraph code in which letters and numbers are represented by strings of dots and dashes (short and long signals).

The young Guglielmo Marconi shortly after his arrival in England, in 1896. In front of him is his apparatus for sending messages by radiotelegraphy.

## The birth of radiotelegraphy

The early development of radio is credited largely to an Italian engineer, Guglielmo Marconi (1874–1937). Unaware of the developments made by Branly, Lodge, and Popov, Marconi's interest in science led him also to begin experimenting with radio, in 1894. In 1895, aged only 21, Marconi used radio to transmit telegraph signals more than a mile (over 1.5 km). The invention became known as radiotelegraphy.

Marconi moved to Britain the following year because the Italian government was not interested in his ideas. On December 12, 1901, he used antennas hung from kites to send a Morse code message from Cornwall in England to St John's, Newfoundland—a distance of 2,200 miles (3,500 km). Although Marconi had little scientific training (and the successful reception of his long-distance Morse code message has been disputed), he won the Nobel Prize for Physics in 1909.

A radio station in Berne, Switzerland, in the 1920s. Signals were sent via the tall transmitter masts.

**Curriculum Context**

Students should be aware that people continue inventing new ways of doing things.

### Voices on the radio

So far, radio was better than the telegraph only because it did not need wires. The telephone, which needs wires, can carry the sounds of speech. Could radio be made to carry the human voice? This question led to the development of radiotelephony. The early work was done by Canadian-born American electrical engineer Reginald Fessenden (1866–1932), who invented modulation. Radiotelegraphy sends out pulses of short and long signals representing the dots and dashes of Morse code. In radiotelephony, the transmitter sends out a continuous signal, called a carrier wave, whose amplitude (strength) is varied (modulated) in step with the variations in the sound signals from a microphone. Fessenden first demonstrated AM (amplitude modulation) in 1903, and by 1906 he had transmitted speech and music from a radio station in Massachusetts.

## Picking up the signal

The new system needed a better detector, however. This came in the form of an improved crystal detector, produced by American electrical engineer Greenleaf Pickard (1877–1956) in 1906. It used a crystal of carborundum (silicon carbide), galena (lead sulfide), or silicon, and worked by rectifying the incoming radio signal, converting it from alternating current (AC) to direct current (DC). The detector connected to the radio circuit by an adjustable thin wire, which soon earned it the nickname "cat's whisker."

English engineer John Fleming (1849–1945) had invented an even better rectifier/detector system in 1904. This was a two-electrode vacuum tube called a diode. Two years later, American engineer Lee De Forest (1873–1961) added a third electrode to make the audion, or triode, vacuum tube, which could be used as an amplifier to boost weak radio signals. With the aid of the new devices, radio engineers could build better circuits for transmitters and receivers. In 1917, Marconi began making VHF (very high frequency) transmissions, although VHF did not come into its own until it was required by television 20 years later. By 1924, Marconi was also sending speech signals from England to Australia using short-wave radio. Vacuum tubes have largely been replaced in radios now by much smaller transistors, invented by three scientists from the Bell Laboratories, New Jersey, in 1948.

### Vacuum tube

An airtight glass tube in which electricity is conducted by electrons passing through a partial vacuum from a cathode to an anode; also called an electron tube.

### Family entertainment

De Forest also arranged the first scheduled radio broadcast, in 1910, which was a performance of the Metropolitan Opera in New York, featuring the famous Italian opera singer Enrico Caruso. Before the invention of television, radio was unrivaled as a means of family entertainment; the period from 1920 to 1950 is often known as the "Golden Age" of radio. Among the many famous radio broadcasts is the announcement—by President Franklin D. Roosevelt on December 9, 1941—that the United States had entered World War II. De Forest's contributions to the the embryonic science of electronics did not end with the triode. He patented more than 300 inventions, including new ways of speeding radio signals, making an optical sound track for talking pictures, and sending messages by radio using facsimile transmission, which today we call fax.

### Better tuning

Radio receivers improved in 1912 when Fessenden devised the heterodyne circuit, which allowed more selective tuning. They improved even further when American engineer Edwin Armstrong (1890–1954) invented the superheterodyne circuit in 1918, which can detect very weak signals. This greatly simplified the tuning of radios (and later, televisions and satellites) to waves of a particular frequency (number of vibrations per second). Previously, tuning had been a complex process involving many dials.

### Frequency modulation

Armstrong's major contribution to radio development came in 1933, when he invented FM (frequency modulation). In this technique, the frequency (not the amplitude) of the transmitted carrier wave is modulated by the broadcast signal. As a result, the transmission is far less sensitive to atmospheric radio disturbances called static, producing a significant increase in the quality of the received sound.

---

**Electronics**

A branch of physics and of electrical engineering that involves the manipulation of voltages and electric currents through the use of various devices for the purpose of performing some useful function.

**Static**

A hissing or crackling noise caused by electrical interference from the atmosphere.

# Radio Transmission and Reception

**1** At the radio station, microphones turn sound waves into electrical waves.

**2** A carrier wave of regular frequency and amplitude is generated by an oscillator. The electrical sound signals are amplified (strengthened) and used to modulate (vary) the carrier wave. In AM radio signals, the amplitude (power) of the carrier wave is varied. In FM radio signals, the frequency (number of vibrations per second) of the carrier wave is modulated.

**3** The carrier waves are amplified and radiated from the station's transmitting antenna.

**4** The antenna of a radio picks up radio waves from many different transmitters, and converts them into electrical signals.

**5** To hear a program, the listener tunes the radio to one carrier frequency, or channel.

**6** The converter (or frequency changer) converts the frequency to a lower one called the intermediate frequency, which is always the same. Amplifiers strengthen this signal.

**7** The demodulator separates the sound signal from the carrier wave. The sound signal can be adjusted by the volume control dial.

**8** Another amplifier strengthens the signal so that it is loud enough to work a loudspeaker, which converts the signals back into sound waves.

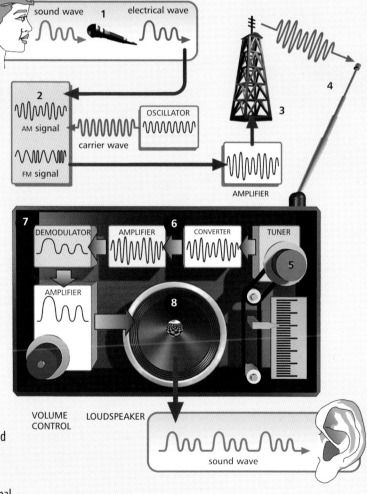

The invention of radio 61

# Elusive electrons

**Every modern science student knows what an electron is and how important it is in understanding electricity and atomic physics. But things were very different 100 years ago, when an English physicist called J.J. Thomson discovered what was then the smallest particle of matter known.**

### Curriculum Context

Students should know that matter is made of minute particles called atoms, and atoms are composed of even smaller components. These components have measurable properties, such as mass and electrical charge.

Until the end of the 19th century, various discoveries in physics had given rise to a host of unanswered questions. Here are just a few: objects can be made to hold a charge of static electricity, but what form does the charge take? In an electric current, charges flow along a conductor, but what are these charges, and are they different from the electrostatic ones? A high voltage across the plates of a vacuum tube produces cathode rays, but what are the rays made of? And if matter is made of atoms, what are atoms made of?

## Finding particles

It turned out that the answers to these questions depended on an invention—the vacuum pump—that allowed scientists to remove nearly all the air from a piece of apparatus. One of the first experimenters to use it was the German glassblower and manufacturer of laboratory equipment Heinrich Geissler (1815–79). In about 1850, he sealed metal plates inside a glass vacuum tube containing only traces of a gas (such as neon or argon). He connected the plates to a source of high-voltage electricity, and obtained pretty lighting effects as the gas glowed in the container.

The Geissler tubes became a popular novelty, but were used for serious experiments by two German physicists: Julius Plücker (1801–68) in 1859, and Johann Hittorf (1824–1914) in 1869. They maintained that the light resulted from "rays" that left the negatively charged plate, or cathode, of the Geissler tube and traveled

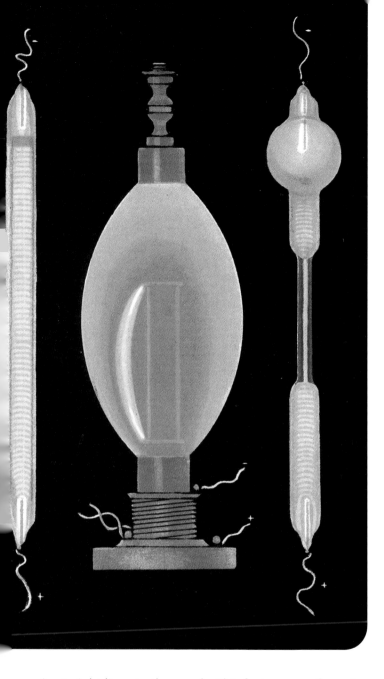

Geissler tubes are glass tubes or bulbs with most of the air pumped out of them. A high voltage applied to metal plates sealed into the ends of the tubes makes the few remaining gas atoms glow as they conduct the electric current between the plates.

in straight lines to the anode. This fact was confirmed in 1879 by English physicist William Crookes (1832–1919), who also suggested that the "rays" might, in fact, be particles. Sixteen years later, French physicist Jean Perrin (1870–1942) deflected these cathode rays using magnetic and electric fields, proving that they are made up of negative electric charges.

### Anode

The positively charged plate or electrode.

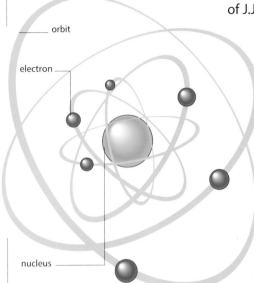

Rutherford's model of the structure of the atom shows electrons orbiting a central nucleus.

The scene was now set for the pioneering experiments of J.J. Thomson (1856–1940). Joseph John Thomson (always known as J.J. Thomson) was an English physicist. He was born near Manchester in the north of England. At the age of 14, he began to train as a railroad engineer. After that he won a scholarship to Cambridge University, graduating in 1880. He went to work in the famous Cavendish Laboratory under John Strutt, Lord Rayleigh (1842–1919), and succeeded him in the professorship in 1884.

## Subatomic "corpuscles"

Thomson also deflected cathode rays with electric and magnetic fields, measured their speed (proving that they travel more slowly than light waves), and figured out the ratio of their charge (e) to their mass (m). This ratio, e/m, was 1,000 times smaller than that for a hydrogen ion (the smallest charged atom), and so Thomson deduced that cathode rays must consist of minute, negatively charged particles. He announced the discovery of these first subatomic particles, which he called "corpuscles," in 1897.

Two years later, Thomson found that the subatomic particles have a mass equal to about one two-thousandth of the mass of a hydrogen atom. Their existence had already been predicted in 1874 by the Irish physicist George Stoney (1826–1911); in 1891, he had named them "electrons." The electron turned out to be the long-sought-after unit of electricity responsible for electrostatic charges. Also, it became clear that a flow of electrons along a conductor constitutes an electric current. And because electrons come from the metal of the cathode in a discharge tube, they must be a fundamental part of all atoms.

### Curriculum Context

Students should know that each atom has negatively charged electrons surrounding a positively charged nucleus.

Thomson went on to study canal rays (positive rays emitted by the anode of a discharge tube). This work led in 1912 to finding a method of separating charged particles. This helped English physicist Francis Aston (1877–1945) to develop the mass spectrograph (a device for finding the elements in a substance) in 1919.

> **Curriculum Context**
>
> The curriculum requires that students understand the structure of the atom.

### Toward the structure of the atom

When Thomson retired in 1919, he was succeeded by his former assistant, New Zealand-born English physicist Ernest Rutherford (1871–1937). Rutherford eventually proposed a structure for the atom that included the atomic nucleus. Seven of Thomson's assistants later won Nobel prizes, an award that he himself received in 1906.

## Electricity Flow

Electrons are basic components of all atoms, in which they normally orbit the nucleus. In most metals—such as the copper wire of an electric cable—many of the electrons wander away from their atoms to form a "sea" of free electrons (bottom section of diagram). When a voltage is applied to the cable (middle section), the free electrons move and become the electric current flowing along the cable. Copper is an example of a metal that is said to have low resistance. A metal with very high resistance, such as tungsten used for elements in electric heaters, has only a few free electrons (top section). As a result, a voltage causes only a small current to flow, and the metal becomes hot.

HIGH-RESISTANCE METAL—
SMALL CURRENT FLOW

LOW-RESISTANCE
METAL—CURRENT
FLOW

LOW-RESISTANCE METAL—
NO CURRENT FLOW

insulation

atom of high-resistance metal

electrons move

free electron

atom of low-resistance metal

Elusive electrons

# The first automobiles

**The automobile was the result of a long period of trial and error in the quest for a motorized road vehicle. The early machines had to use a steam engine, the only motive power available at the time. But then the internal combustion engine, fueled first by gasoline and later also by diesel, revolutionized this form of travel.**

In 1770, French military engineer Nicolas-Joseph Cugnot (1725–1804) built his second three-wheel gun carriage. It was powered by a two-cylinder steam engine mounted on the single front wheel. It achieved a speed of 3 miles per hour (5 km/h), and was involved in the world's first motor accident when it demolished a wall. German engineer Charles Dietz constructed another extraordinary three-wheel machine in 1835. It had a pair of rocking cylinders that worked cranks to move a chain drive to the rear wheels.

> **Curriculum Context**
>
> The curriculum requires that students describe how and why technology evolves.

Experiments with steam vehicles continued, aimed at producing a tractor or a multipassenger carriage—a bus—rather than a form of personal transportation. In England, Scottish engineer William Murdock (1754–1839) ran a model steam-powered road vehicle in 1784, and in 1789 American inventor Oliver Evans (1755–1819) fitted one of his high-pressure engines to a four-wheel land vehicle. Over in England, engineer Richard Trevithick (1771–1833) built a similar vehicle in 1801. It had large driving wheels, with front wheels that were steered independently, and it could reach a speed of 10 miles per hour (16 km/h). In 1829, English inventor Goldsworthy Gurney (1793–1875) began a regular steam coach service from London to Bath that ran at an average speed of 15 miles per hour (24 km/h).

A scene from an automobile race from Paris to Rouen in 1894. The design of early vehicles was still based on the carriage—hence the name "horseless carriages."

New Yorker Richard Dudgeon built a lightweight steam carriage in about 1865, and in 1873 French engineer Amédée Bollée (1844–1917) demonstrated a 12-seater

The first automobiles

> **Curriculum Context**
>
> Students should understand that chemical reactions occur all around us—for example, in automobile engines.

carriage called *L'Obéissante* (the Obedient One). His *La Mancelle* (The Young Lady from Le Mans) of 1878 had a front-mounted engine driving the rear wheels, and could travel at up to 25 miles per hour (40 km/h). But just as they were starting to become efficient means of transport, steam carriages fell into decline in the face of competition from the railroads.

## Gasoline-powered cars

The next step forward in the development of the automobile came with the work of engineers Karl Benz (1844–1929) and Gottlieb Daimler (1834–1900) in Germany. They realized the potential of the new gasoline engine as a means of powering road vehicles. Benz's first car, a three-wheeler called the *Motorwagen*, dates from 1885. It had a 1-horsepower engine and a maximum speed of 8 miles per hour (13 km/h).

Daimler built his first car a year later, mounting his gasoline engine in a heavier four-wheel vehicle. At first he concentrated on producing engines for other makers' cars, and by 1889 he had a powerful, reliable 3.5-horsepower unit. By 1891, French engineers René Panhard (1841–1908) and Émile Levassor (d. 1897) were emulating Daimler, and building cars with a chassis. The vehicles, known as *système Panhard* (Panhard system), had front-mounted Daimler

The very first cars had only three wheels, but engineers soon favored the stability of four wheels. By the time of the Model T, tiller steering had been replaced by a steering wheel.

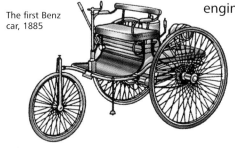

The first Benz car, 1885

Daimler, 1886

Panhard and Levassor, 1894

A 1918 Model T Ford. Unusually, not in the obligatory black livery dictated by Henry Ford.

engines to drive the rear wheels. They had modern Ackermann (double-pivoting) steering, a gear box, and a friction clutch. By 1893, Benz was also manufacturing the more stable four-wheel cars with 3-horsepower engines. The same year saw the first gasoline-engined car built in the United States, by inventor Charles Duryea (1861–1938) and his brother Frank (1869–1967). The first U.S.-manufactured automobile went on sale in 1896. One year later, there was a brief reversion to steam power when the Stanley brothers of Massachusetts—Francis (1849–1918) and Freelan (1849–1940)—launched the Stanley Steamer.

### Quicker, cheaper cars

Soon after the turn of the 20th century, American industrialist Henry Ford (1863–1947) revolutionized car manufacture by introducing mass-production techniques. As a new chassis moved slowly along an assembly line, workers added the engine, transmission, wheels, and finally, the body. In 1908, the method produced the Model T, or "Tin Lizzy," as it was affectionately known by millions of Americans. "You can have it in any color as long as it's black," declared Ford. The motor age was born.

> **Curriculum Context**
>
> The curriculum requires that students know the standard units of measure of energy, power, and transportation.

> **Curriculum Context**
>
> Students should be able to distinguish between continuous, intermittent, custom, and other manufacturing systems.

# Airships

**Only weeks after the first passengers flew in a Montgolfier hot-air balloon in 1783, another type of balloon soared over the rooftops of Paris. It was filled with hydrogen gas, and soon hydrogen balloons became the predominant lighter-than-air craft.**

Controlled flight in a balloon was not at first a possibility because there was no power source small enough and light enough to fit in a balloon's basket. Balloons were also at the mercy of the winds, forced to travel wherever the weather took them unless they were tethered to the ground. The advantages that could be gained by having a free-floating, controllable balloon were obvious. As the 19th century came to a close, every military power that could afford such a program was rushing to build a "ship of the air."

French physicist Jacques Charles (1746–1823) was the first to fill a balloon with hydrogen. However, these early balloons could carry just a few people and only in the direction the wind took them. The question was: how could they be made to carry more, and in a predetermined direction? The answer was to make the balloon bigger and to add a motor and a rudder.

### Hydrogen

A colorless, highly flammable gaseous element (atomic number 1), the lightest of all gases, and the most abundant element in the universe; chemical symbol H.

Considering the pioneering efforts of the French, it is not really so surprising that a Frenchman, Henri Giffard (1825–82), was the first person to create such an aircraft.

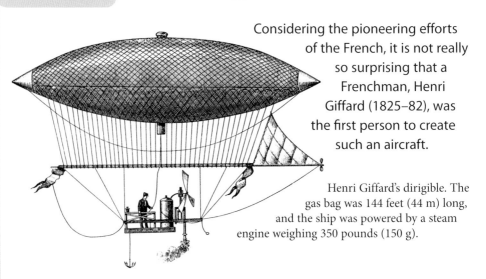

Henri Giffard's dirigible. The gas bag was 144 feet (44 m) long, and the ship was powered by a steam engine weighing 350 pounds (150 g).

This postcard depicts Brazilian aviator Alberto Santos-Dumont's airship, in which he won a prize of 100,000 francs for flying it from Saint Cloud to the Eiffel Tower (shown) and back on October 19, 1901: a distance of 6.8 miles (11 km).

## A steerable airship

The resulting craft, built in 1852 and now known as an airship, was originally called by the French name "dirigible," meaning "steerable". Instead of a normal balloon-shaped bag to hold the hydrogen, Giffard's craft consisted of a long, cigar-shaped envelope, or gasbag. This was draped with a network of ropes supporting an open cabin, or gondola, that carried the pilot and the engine. The envelope kept its shape because of the pressure of the gas inside—an arrangement that defines it as a nonrigid airship.

## Lacking in power

It was powered by a 3-horsepower coke-fired steam engine that rotated a two-bladed propeller. However, the weight of this engine meant that a great deal of gas was needed to lift the ship off the ground. A rudder at the rear of the gondola steered the craft. The gondola was suspended well below the gasbag in case sparks from the engine's boiler ignited the highly

### Curriculum Context

Students should be aware that a technology often advances with the introduction of new, unrelated technologies.

flammable hydrogen. Unfortunately for Giffard, his craft was fated always to be underpowered. Although his dirigible was an improvement on the normal balloon, any aircraft with a top speed of 6 miles per hour (9.5 km/h) was always going to be uncontrollable in wind speeds of 7 miles per hour (11 km/h) or more.

### Electric motors and gasoline engines

On Giffard's first trip, he flew from Paris to a village a distance of 17 miles (27 km) away, at an average speed of 5 miles per hour (8 km/h). But to fly in anything other than a light breeze, a more powerful engine would be needed, such as an electric motor. In 1883, French brothers Albert Tissandier (1839–1906) and Gaston Tissandier (1843–99) became the first people to power a dirigible with an electric motor. That was also the method used in 1884 by French army engineers Charles Renard (1847–1905) and Arthur Krebs (1847–1935) for their airship *La France*. The battery-powered motor generated nearly 9 horsepower, and the inventors successfully flew the 164-foot (50-m) airship around a circular course near Paris at 14 miles per hour (23 km/h).

However, back in 1872 German engineer Paul Haenlein (1835–1905) had fitted the recently invented (and much lighter) internal-combustion engine to a dirigible. To save even more weight, Haenlein used hydrogen from the gas bag to power the engine (although this did reduce the distance the craft could travel before it began to lose height). Inventors began increasingly to build airships powered by gasoline engines.

### Semirigid and rigid airships

The semirigid design evolved in 1898: metal frames in the nose and rear were connected by a wooden lattice keel. The first rigid airship, which had an overall internal metal framework to maintain its shape, was built by Austrian inventor David Schwartz (1852–97) in 1897.

> **Curriculum Context**
>
> Students are expected to distinguish in transportation activities between the various power systems.

A zeppelin is caught in searchlights in a night raid over Britain during World War I.

Aluminum was now becoming the favored metal for construction. It was used by German engineer Graf (Count) Ferdinand von Zeppelin (1838–1917) for his first airship, the *LZ-1*, in 1900. The 420-foot- (128-m-) long craft was not a great success, however. The *LZ-2* followed in 1905 and the *LZ-3* in 1906, but the first truly successful zeppelin was the *LZ-4* of 1908. At the time the *LZ-4* was the largest airship in the skies, at 446 feet (136 m) long. On July 4, 1908, the *LZ-4* traveled at 40 miles per hour (60 km/h) for 12 hours over Switzerland.

# Airships

Nonrigid airships (or blimps) have no internal framework to support the shape of the outer envelope, which is filled with gas and one or more air-filled ballonets. Air can be bled from the ballonets or pumped in to increase or decrease lift. Rigid airships (or zeppelins) have an internal framework that supports the outer envelope.

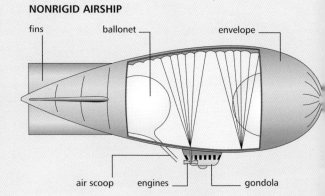

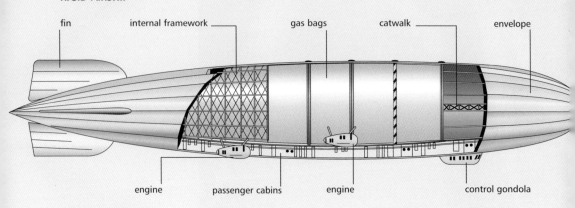

**Curriculum Context**

The curriculum requires a knowledge of the interrelationships between temperature, pressure, and volume of gases contained within a closed system.

The airships were getting bigger, with two, three, and eventually four engines. The 486-foot (148-m) *Deutschland* inaugurated the world's first commercial airline in 1910. It was followed by *Graf Zeppelin*, a 771-foot (235-m) airship that could carry passengers across the Atlantic Ocean at speeds of up to 81 miles per hour (130 km/h).

Between 1910 and the outbreak of World War I (1914–18) in 1914, over 34,000 people had their first taste of air travel in one of Zeppelin's airships. World War I proved to be the real spur to airship construction.

Germany led the world in airship design and building at this time, constructing 88 military airships between 1914 and 1918. London had the dubious honor of being the first city to be attacked from the air, when a group of what had come to be known as "zeppelins" made a nighttime raid on Britain's capital.

## The demise of the airship

Between 1920 and 1933, the U.S. navy built five airships. Three of them crashed: the *Shenandoah* (1925), the *Akron* (1933), and the *Macon* (1935). In 1936 Germany's ill-fated *Hindenburg* made its first flight, and in the same year it began regular transatlantic crossings. Unfortunately for all concerned, it proved to be the final chapter in the history of the airship. Although hydrogen is the lightest of all gases, it is also one of the most flammable. In 1937, when the *Hindenburg* was approaching its mooring at Lakehurst, New Jersey, its hydrogen caught fire and exploded, killing a total of 36 passengers and crew. The 787-foot (240-m) British airship *R 101* had suffered a similar fate in 1930, when it crashed in France on its way to India. Of the 54 people on board, 48 were killed.

These disasters signaled the end of the hydrogen-filled airship. Never again would airships of this magnitude be built. Within two years, their role as passenger craft had been taken over by faster, cheaper airplanes. The U.S. navy used airships for antisubmarine patrols until 1962, however, and has recently reconsidered using airships to counter the threat of low-flying cruise missiles. There were also plans for cargo-carrying airships, but this role was taken over by helicopters.

The airship is now little more than a novelty item, filled with the slightly heavier but much safer gas helium, and used for pleasure flights and for advertising—most famously, the Goodyear Tire and Rubber Company's three blimps.

> **Curriculum Context**
>
> Students should be able to describe the factors that affect the adoption or rejection of technology.

> **Helium**
>
> A light, colorless, nonflammable inert gaseous element (atomic number 2) occurring in natural gas, in radioactive ores, and in small amounts in the atmosphere; chemical symbol He.

# Airplanes

**Even before the invention of lighter-than-air flying machines, such as balloons and airships, people wanted to imitate birds and take to the air. For this reason many of the early designs, such as those drawn by Leonardo da Vinci in about 1500, had flapping wings. It was almost another 500 years before this ambition was realized.**

A flapping-wing machine is called an ornithopter. No such machine was ever built except in model form; nor would a full-size version have worked even if built, because it depended on human muscle power for propulsion. (Human-powered aircraft have been flown in recent times, but aided by modern knowledge of aerodynamics, mechanisms, and materials.)

The first heavier-than-air machines to fly were kites, invented by the Chinese in about 1000 B.C. In the late 19th century, human-carrying kites were built, including one designed for military use by English soldier Baden Baden-Powell (1860–1937) in 1894, and improved in 1901 by American showman "Colonel" Samuel Cody (1867–1913). Today's ski-kites and microlight aircraft continue the tradition. But real progress in this area was not achieved until people began experimenting with gliders.

### Pioneer gliders

One of the first to do so was English inventor George Cayley (1773–1857). In 1808, he flew an unmanned glider with a wing area of about 320 square feet (30 sq m). Then, in 1853, he built a 300-pound (135-kg) man-carrying glider. The passenger was Cayley's footman—unaware that he was the first person ever to fly in a heavier-than-air machine. Three years later, French naval officer Jean-Marie Le Bris (1817–72) made a short gliding flight on a beach in northern France. In 1895, Scottish aviator Percy Pilcher (1866–99) made a

> **Kite**
>
> A light framework covered with cloth, plastic, or paper, designed to be flown in the air at the end of a long string.

> **Glider**
>
> A light, unpowered aircraft designed to glide after being towed into the air or launched from a catapult.

Otto Lilienthal flying one of his gliders. The pilot hung beneath the wings, as in a modern-type hang glider.

part-controlled gliding flight in his hang glider, the *Bat*. A year later he built his fourth glider, the *Hawk*. It was steered by a tiller attached to a four-vaned rudder. In 1897, he broke the world record for flight when he flew 820 feet (250 m) in the *Hawk*. Pilcher died in 1899 from injuries sustained when the *Hawk* crashed.

The first person to make a careful study of gliders, and build a steerable one that could be controlled in flight, was German aeronautical pioneer Otto Lilienthal (1848–96). His first manned flight was made in 1891. His early machines copied birds' wings, but later he added a tail for stability, and came up with the idea of

The Wright brothers' plane landing in Le Mans, France, in 1908, piloted by Wilbur Wright (inset).

**Monoplane**

An airplane with only one pair of wings.

two pairs of wings, one above the other. The two-wing arrangement, later called a biplane, remained a feature of nearly all early flying machines. In 1893, Lilienthal tried birdlike jointed wings, but they failed, and in 1896 he became another glider pilot to crash to his death. In the United States, Wilbur Wright (1867–1912) and his brother Orville (1871–1948) read about Lilienthal's pioneering work, which would influence their experiments. Also in the United States, French-born American inventor Octave Chanute (1832–1910) constructed an extremely stable biplane glider in 1896, which was noted by the Wright brothers. By 1903, they had perfected human-carrying gliders of their own.

### Powering the airframe

By the early 20th century, engineers had what we would now call an "airframe." Now all they needed was a suitable power source. Then, the steam engine was the only possibility. As early as 1842, the English-born American engineer William Henson (1812–88) took out a patent for a monoplane aircraft with a steam engine, pusher propellers, and a cabin for several passengers. But he was only ever able to build a nonfunctional model.

In 1848, English inventor John Stringfellow (1799–1883) also made a steam-powered model. It flew a short distance but soon succumbed to the force of gravity, and crashed. In France in 1890, engineer Clément Ader (1841–1926) built *Eole*, a full-sized steam-powered monoplane. It took off under its own power and flew for about 165 feet (50 m), but Ader had no means of controlling the plane.

# The Wright *Flyer*

The Wright brothers' first successful airplane, the *Flyer* of 1903, was based on one of their earlier biplane gliders. They constructed a 12-horsepower lightweight aluminum gasoline engine weighing only 88 pounds (40 kg), and mounted it on the lower wing. It powered two "pusher" propellers that rotated in opposite directions to keep the turning force of torque from trying to rotate the whole airplane. The pilot steered using a combination of the rudder and a technique called "wing warping." This involved flexing the ends of the wings (like a bird) in order to bank the plane into a turn. He used the elevator to make it climb and descend. After several flights, with the brothers taking turns as pilot, the plane was damaged and never flew again. The larger *Flyer II* of 1905 flew around a 24-mile (38-km) circular course in 38 minutes.

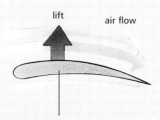

The *Flyer's* wings produced lift because of their shape. Air flowing over the top of the wing has farther to travel and so moves faster than the air below. The faster air moves, the lower its pressure. This causes low pressure above the wing and high pressure below it, producing lift.

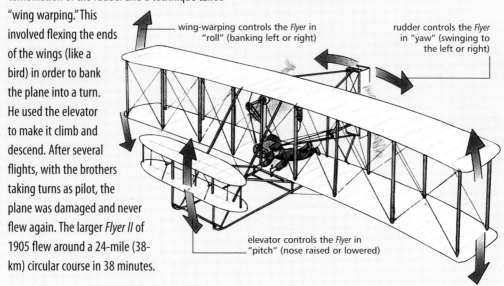

wing-warping controls the *Flyer* in "roll" (banking left or right)

rudder controls the *Flyer* in "yaw" (swinging to the left or right)

elevator controls the *Flyer* in "pitch" (nose raised or lowered)

Another short-lived success was a huge machine built by American-born English inventor Hiram Maxim (1840–1916). His biplane of 1894 had two steam engines, each one driving its own propeller. Launched along a set of rails, it managed to rise less than 3 feet (1 m) before falling to the ground. Two years later, the American astronomer Samuel Langley (1834–1906) built a large, steam-powered model airplane. It flew for about 90 seconds, during which time it covered a distance of 875 yards (800 m).

Steam engines were just too heavy for the task. The obvious alternative was the gasoline engine. In the fall of 1903, Langley built a full-sized airplane with a gasoline engine. Two attempts to launch it from a barge in Washington, D.C. failed—each time, the machine crashed into the Potomac River.

## The Wright brothers

The work of these early pioneers came to the attention of the two Wright brothers in Dayton, Ohio. Having pored over the earlier work of previous pioneers, bicyclemakers and printers Wilbur and Orville Wright set about designing controllable gliders. From watching buzzards in flight, the brothers knew that a successful aircraft would have to be able to bank to one side or another, climb or descend, and turn from left to right. The last glider the Wrights built, in 1902, could perform all these movements. They had also built a wind tunnel to test that the glider's wing and surface shapes were aerodynamic.

Now all that remained was to add a lightweight engine and effective propellers so it could power its own takeoff, flight, and landing. The brothers designed and built a small gasoline engine, and connected it to a pair of propellers (also designed by the brothers) with bicycle chains. Their glider was now an airplane.

On December 17, 1903, Orville Wright took off from level ground in the first powered airplane—*Flyer I* (popularly known as *Kitty Hawk*). The Wright brothers continued to experiment with airplanes throughout 1904, making a number of improvements in the way the original aircraft handled. In 1905, they built *Flyer III*—the world's first practical airplane. It could turn, bank, circle, fly figure-eights, and remain airborne for more than half an hour. The Wright brothers then refused to fly again until they got financial backing from the government or a private company.

> **Aerodynamic**
>
> Designed to reduce or minimize the resistance to air flow as an object moves through it, or by wind that strikes and flows around an object.

> **Curriculum Context**
>
> Students should understand that the impact of technological advances can be far-reaching.

Louis Blériot over the cliffs of Dover, England, in 1909, having just flown across the English Channel from France. In so doing, he completed the first crossing over a large body of water.

At last, in 1908, the U.S. government issued a contract to the Wright brothers for an airplane that could carry a pilot and an observer for 125 miles (200 km) at a speed of no less than 40 miles per hour (65 km/h). With funding from the military, the two brothers were able to deliver such an aircraft a year later. Their plane was a remarkable improvement in the technology that just six years earlier was capable only of carrying a single aviator for 170 ft (51.5 m). Within a year, almost every major army in the world was equipped with airplanes.

The next advances mainly involved materials. Steel and other alloys replaced wood for airframes, and aluminum panels instead of varnished cloth were used to cover them. Jets superseded gasoline engines, and just 44 years after the Wright brothers' first flight, an airplane flew faster than the speed of sound.

### Speed of sound

The speed at which sound travels in a given medium under specified conditions; at sea level, it is 760 miles per hour (1,220 km/h).

# Capturing sound

**Sound recording is the method whereby sound waves are captured on a device in order that they can be reproduced at a later stage. The digital systems employed to achieve this today are just the latest innovation in a process that began over 130 years ago.**

The age of sound recording arrived in 1877, when the U.S. inventor Thomas Edison (1847–1931) launched his mechanical phonograph cylinder. It was the first commercially available machine that could capture sound, and then play it back.

**Curriculum Context**

Students are expected to research and describe the contributions of scientists and inventors.

Edison's machine was based on experiments carried out in the 1850s. Scientists interested in sound attached a flexible diaphragm (thin, flat disk or cone) to a fine needle that rested against a moving, soot-covered glass plate. Sound made the diaphragm vibrate, causing the needle to trace wavy lines in the soot, showing that sound travels in waves. Edison attached a speaker horn to a needle resting on a tinfoil drum. The voice was recorded by turning the drum by hand while speaking into the speaker horn. The needle made a series of tiny indentations in the tinfoil that represented the sound waves. In order to play the sound back, the needle was replaced with a bristle, and the drum rotated.

**Patent**

A grant made by a government that confers upon the creator of an invention the sole right to make, use, and sell that invention for a set period of time.

### The gramophone

Probably best known for developing the telephone, Alexander Graham Bell (1847–1922), a Scotsman who settled in the United States, had developed an interest in sound through his work with people suffering from deafness. He wanted to improve the sound quality of the phonograph, and in 1886 he took out a patent on his "graphophone". Bell's machine differed from Edison's mainly in that the needle vibrated from side to side, rather than up and down.

Thomas Edison's first phonograph. It was patented in 1878.

A year later, the German inventor Emile Berliner (1851–1929) demonstrated his own gramophone, a machine that recorded sound in grooves on a thin disk of zinc covered in wax. He produced a metal mold of a master disk and realized he could press cheap copies of recordings. By 1901, Berliner's Victor Talking Machine Company had a catalog of over 5,000 recordings for the gramophone. The mass production of disks meant that people could hear their favorite artists in the comfort of their own homes.

## Magnetic recordings

While Edison and Bell were developing mechanical systems, other inventors were capturing sound using magnetism and electricity. In 1878, American Oberlin Smith (1840–1926) recorded the electrical signals produced by a telephone onto a steel wire. However, neither industry nor the public seemed interested in his new form of recording, and in 1888 Smith donated his ideas to the public by publishing them in the journal *Electrical World*.

Ten years later, Danish physicist Valdeman Poulsen (1869–1942) created his telegraphone, the first dictation and telephone answering machine. The telegraphone recorded messages onto a steel wire, which was coiled around a cylinder that rotated under an electromagnet connected to a telephone earpiece. However, it was not popular.

> **Curriculum Context**
>
> The curriculum requires that students are aware that scientific knowledge is made public through publications in scientific journals.

**Curriculum Context**

Students should be aware that technological knowledge is often not made public because of patents, and to safeguard the financial potential of the idea or invention.

In 1888, Oberlin Smith put forward the idea that strips of fabric covered with iron filings could be used to record sounds. However, it was not until 40 years later that German scientist Fritz Pfleumer patented the first magnetic recording tape, consisting of a paper strip coated with a magnetizable steel layer. The German electrical manufacturing company AEG bought the patent, and developed it by coating plastic tape with iron-oxide powder. In 1935, AEG demonstrated its new recording machine—which was the first reel-to-reel tape recorder, the magnetophon—by recording the London Philharmonic Orchestra.

These older tape recorders often needed yards of tape to record the sound. Gradually, however, tapes became smaller, and in 1964 Philips developed the cassette tape. This cassette, which was only 4 inches (10 cm) long, was very popular.

### Improving sound quality

By the 1920s, the issue of sound quality had became very important. In 1925, Bell Laboratories perfected a system of recording sound electrically. By converting sound into electric current, they could replace the huge recording horns with smaller devices: microphones. Electrical microphones enabled all of the recording to be heard, whereas in mechanical recordings the louder sounds always came out best. The microphone allowed softer sounds, such as those produced by violins and harps, to be recorded and clearly heard.

**Curriculum Context**

Students should know that electricity in circuits can produce sound and magnetic effects, and vice versa.

Early recording machines could not distinguish between sounds coming from different directions. British physicist Alan Dower Blumlein (1903–42) led research at EMI (Electrical and Musical Industries) to make recordings sound more authentic. The first stereo records were eventually introduced in 1958, and were made by recording two channels of sound to reproduce the direction and space of a live concert.

# Microphones and Loudspeakers

Microphones and loudspeakers are examples of transducers—devices that convert energy from one form to another. Sound waves carry energy through the air as a regular pattern of disturbance in the air pressure. A microphone can convert this sound energy into electrical energy. The sound waves push against a piece of plastic called a membrane or diaphragm, which in turn exerts pressure on a piezoelectric crystal. When a force is applied to crystals of this type, they produce an electric charge. The charge flows through a circuit inside the microphone as an electric current, whose size depends on the amount of energy in the original sound waves.

In a loudspeaker the opposite occurs. An electric current generated by a radio, stereo, or TV is used to power a coil of wire, generating a magnetic field. The coil is placed between the poles of a large, permanent magnet, where it vibrates as the current flows through it. A large paper cone attached to the coil also vibrates, producing the variations in air pressure that we hear as sounds.

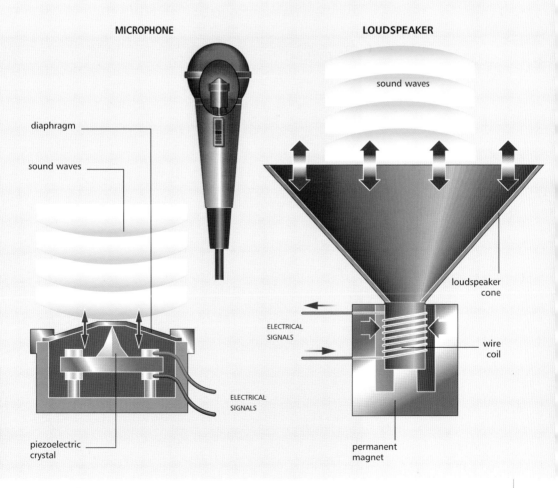

Capturing sound

# Plastics

**Plastic are synthetic (human-made) materials that can be easily molded into a variety of shapes, usually by applying heat or pressure. Cheaper than many other materials, plastics have numerous uses in the home, in industry, and in medicine.**

Before the invention of synthetic plastics, natural materials, often those with plasticlike properties, were widely used instead. Combs and other intricately shaped objects were made from carved animal horn. Gutta-percha, a tree gum from Malaysia, was used for electrical equipment and cables, and shellac, an insect secretion, was used to mold records.

A radio with a Bakelite cabinet. Because Bakelite was inexpensive and also nonconductive, it was often used for electrical equipment, such as radios and television casings.

It was nitrocellulose, produced by dipping cotton into a mixture of nitric and sulfuric acids, that formed the basis of the first plastics. In 1855 Alexander Parkes, (1813–90), a British chemist, created a flexible, durable material called Parkesine from a mixture of nitrocellulose, alcohol, camphor, and vegetable oils.

## Celluloid

In the United States, John Wesley Hyatt, who was looking for a cheap substitute for ivory to make billiard balls, heard of Parkes's work. Hyatt developed a new plastic called celluloid, and in 1872 set up a company to manufacture the substance. Celluloid was used for making collars and cuffs, knife handles, photographic films, and billiard balls.

> **Curriculum Context**
>
> Students should be able to evaluate the impact of technical advances on society.

### Bakelite

Bakelite, the first truly synthetic plastic, came onto the U.S. market in 1909. Belgian chemist Leo Hendrik Baekeland (1863–1944) invented the substance while searching for a synthetic alternative to rubber, which tended to dry out and crack. Made from phenol and formaldehyde, Bakelite is tough and easy to produce. The dark-colored plastic was used in many household appliances, and as an insulator in electrical equipment.

### Synthetic garments

In 1883, Joseph Swan patented a process to produce thin nitrocellulose filaments while researching light bulb technology. French chemist Count Hilaire de Chardonnet (1839–1924) developed these fibers for use in clothing. From 1935, artificial fabrics produced from fibers of this type were called rayon. In 1928, Wallace Hume Carothers (1896–1937) was hired to lead a research team investigating the production of polymers—synthetic plastics formed by joining molecules into long chains. After many tests, he produced a material named 6,6-polyamide—later called nylon—a fiber that was elastic, tough, and water, resistant. However, it was not until 1940 that the first nylon stockings appeared in stores.

### Polyethene

In 1933, while experimenting with reactions under very high pressure, Reginald Gibson and Eric Fawcett, both working for the British company ICI, set up a reaction between ethene and benzaldehyde (both obtained from crude oil) under pressure 1,700 times greater than that at sea level. A white, waxy product was formed. This later turned out to be polyethene. Further experiments caused explosions, and it was not until 1937 that the reaction was fully understood and commercialized. Polyethene is now used to make plastic bottles, fibers, photographic film, and many other everyday products.

---

**Curriculum Context**

The curriculum requires students to know that physical and chemical properties influence the development and application of everyday materials such as plastics.

**Curriculum Context**

Students should be able to describe the arrangement of atoms in polymers.

**Curriculum Context**

Students should be able to describe the chemical, mechanical, and physical properties of manufacturing material such as plastics.

# Making Polymers

Polymers are giant molecules made up of many smaller parts (called monomers) joined in a long chain. Natural polymers include lignin, the main component of wood, and collagen, the substance that forms our hair and fingernails. Synthetic polymers, or plastics, are among the most useful and widespread of modern materials. By altering the structure of the building blocks, and the conditions under which they react, chemists can make polymers with properties to suit particular needs—from clothing to aircraft parts.

A polyethene molecule, a typical homopolymer (made up of only one type of monomer), is a straight chain of between 1,000 and 5,000 ethene molecules. Polyethene is made from ethene gas, which in the modern production process is pressurized and heated to 300°F (150°C) in the presence of titanium and aluminum catalysts. When an ethene molecule passes close to the surface of the catalyst, one of its bonds breaks (1), and it forms a new bond with the end of the polymer molecule (2). In this way, the polymer grows from the surface of the catalyst (3). Polyethene molecules are long, straight chains that pack closely together, producing a dense, tough material. Some polymers have branched chains that do not pack closely together, making a less dense plastic. Bonds called

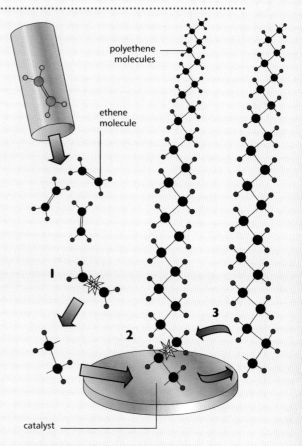

crosslinks can also be formed between chains. Polymers with no crosslinks are called thermoplastics and can be reshaped when they are heated. Thermosetting polymers have many crosslinks and harden permanently once they have been molded.

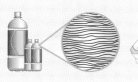
STRAIGHT CHAIN

BRANCHED CHAIN

CROSSLINKED

# Timelines 1826–1835

## Technology

**1826** French chemist Joseph Niépce (1765–1833) takes the first photograph (on a metal plate).

**1828** Scottish engineer James Neilson (1792–1865) invents the hot-air process for smelting iron.

**1829** The Rocket steam locomotive, built by English engineer George Stephenson (1781–1848), wins the Rainhill Trials—a competition to find motive power for the Liverpool & Manchester Railway, which opens a year later.

**1829** French teacher Louis Braille (1809–52), blind himself from the age of three, invents the Braille alphabet to enable blind people to read using touch.

**1830** American engineer Peter Cooper (1791–1883) builds Tom Thumb, the first railroad locomotive to be made in the U.S.

**1831** English scientist Michael Faraday (1791–1867) makes a simple dynamo.

**1831** American physicist Joseph Henry (1797–1878) invents the electric bell. He also builds and operates an experimental telegraph over a distance of 1 mile (1.6 km).

**1832** English physicist Charles Wheatstone (1802–75) invents the stereoscope, which produces an image in three dimensions from a pair of stereo photographs.

## Biology and Medicine

**1827** American ornithologist and artist John James Audubon (1785–1851) publishes the first part of *Birds of America*.

**1828** French physiologist Pierre Flourens (1794–1867) explains how the semicircular canals in the inner ear control the sense of balance.

**1828** Estonian naturalist Karl von Baer (1792–1876) founds the science of embryology, the study of embryos.

**1829** German anatomist Martin Rathke (1793–1860) finds evidence of gill structures in the embryos of birds and mammals.

## Physical Sciences and Math

**1826** Swiss-born French mathematician Jacques Sturm (1803–55) determines the speed of sound in water, and finds it to be much faster than in air.

**1826** French physiologist René Dutrochet (1776–1847) describes the phenomenon of osmosis, in which a solvent passes through a semipermeable membrane separating solutions of different concentrations.

**1827** French mathematician Joseph Fourier (1768–1830) suggests that world climate is affected by human activities—a view that has gained favor again in recent times.

**1827** German physicist Georg Ohm (1789–1854) publishes Ohm's law, which states that the voltage across a conductor divided by the current flowing through it is a constant, called the resistance.

**1827** Scottish botanist Robert Brown (1773–1858) discovers Brownian motion, the continuous random movement of microscopic particles suspended in a liquid.

**1827** German physicist Karl Gauss (1777–1855) develops differential geometry.

**1829** Scottish chemist Thomas Graham (1805–69) formulates Graham's law.

The Ages of Steam and Electricity

## Technology

**1832** Belgian physicist Joseph Plateau (1801–83) invents the stroboscope—a rapidly flashing light that can make rotating machinery appear stationary.

**1833** German physicists Karl Gauss (1777–1855) and Wilhelm Weber (1804–91) demonstrate a needle telegraph (the arrival of a signal is indicated by movement of a magnetic needle).

**1833** English mathematician Charles Babbage (1792–1871) begins work on his "analytical engine," a type of mechanical computer, but the work is never completed.

**1834** American engineer Cyrus McCormick (1809–84) patents a reaping and binding machine, a forerunner of the combine harvester.

**1834** English phonographer Isaac Pitman (1813–97) invents the system of shorthand writing that bears his name. It is soon adopted throughout the English-speaking world.

**1835** American physicist Joseph Henry (1797–1878) invents the electromagnetic relay.

**1835** American manufacturer Samuel Colt (1814–62) produces a revolver with interchangeable parts.

## Biology and Medicine

**1831** Scottish botanist Robert Brown (1773–1858) describes the nucleus of a cell (in this case a plant cell).

**1833** American army surgeon William Beaumont (1785–1853) explains the role of gastric juices in digestion.

**1834** French agricultural chemist Jean Boussingault (1802–87) discovers nitrogen fixation in plants.

**1835** English physician James Paget (1814–99) discovers the trichina parasite, a roundworm that can be contracted by humans by eating undercooked pork.

## Physical Sciences and Math

**1830** Scottish geologist Charles Lyell (1797–1875) begins publication of his most influential book, *The Principles of Geology*.

**1830** English mathematician George Peacock (1791–1858) first puts forward the laws of numbers in his book *Treatise on Algebra*.

**1830** Scottish writer Mary Somerville (1780–1872) publishes *The Mechanism of the Heavens*, a popularization of *La Mécanique Céleste* by French astronomer Pierre-Simon de Laplace (1749–1827).

**1831** English explorer and naval officer James Ross (1800–62) locates the position of the north magnetic pole (which changes continuously).

**c.1832** German physicist Karl Gauss (1777–1855) formulates Gauss's law.

**1832** French chemist Pierre Robiquet (1780–1840) discovers codeine (in opium from poppies).

**1834** German chemist Justus von Liebig (1803–73) synthesizes melamine, the basis of a range of modern plastics.

**1834** English scientist Michael Faraday (1791–1867) formulates the laws of electrolysis. Electrolysis is the process by which an electrical current running through an electrolyte causes chemical reactions to occur at the electrodes.

# Timelines 1836–1844

## Technology

**1836** English physicist William Sturgeon (1783–1850) invents the moving-coil galvanometer, a sensitive instrument for detecting electric currents.

**1836** American engineer Thomas Davenport (1802–51) makes a model streetcar.

**1837** German-born Russian printer Moritz von Jacobi (1801–74) invents a form of electrotyping in which a whole page of type is made by electroplating a mold taken from a page of type set in the usual way.

**1837** English physicists William Cooke (1806–79) and Charles Wheatstone (1802–75) patent a five-needle electric telegraph.

**1838** American inventor Samuel Morse (1791–1872) demonstrates a single-wire electric telegraph over a 10-mile (16-km) circuit at New York University. The electromagnetic receiver "clicks" when telegraph signals arrive.

**1838** The paddle steamer *Sirius* is the first steamship to cross the Atlantic Ocean under steam power alone (taking 18 days). Only hours later SS *Great Western* arrives, having taken four days less.

**1839** French painter and physicist Louis Daguerre (1787–1851) invents the daguerreotype, a type of photograph taken on metal plates (with the image reversed left to right).

## Biology and Medicine

**1836** American botanist Asa Gray (1810–88) draws up a method for classifying plants based on their fruits and seeds.

**1836** English naturalist Charles Darwin (1809–82) completes his epic voyage on HMS *Beagle*; his observations during the voyage eventually enable him to propose his theory of evolution by natural selection.

**1837** French physiologist René Dutrochet (1776–1847) describes the role of chlorophyll in photosynthesis—as the substance that absorbs sunlight and converts its energy to food.

**1839** German biologist Theodor Schwann (1810–82) postulates that all living matter is made up of cells.

## Physical Sciences and Math

**1836** English chemist John Daniell (1790–1845) invents the Daniell cell, a primary cell (battery) with zinc and copper electrodes.

**1836** French chemist Alexandre Becquerel (1820–91) identifies the photovoltaic effect (whereby light produces an electric current), the principle of most modern photocells.

**1836** Swedish chemist Jöns Berzelius (1779–1848) discovers catalysts—substances that accelerate or slow down chemical reactions without taking part in them.

**1836** English chemist Edward Davy (1806–85) discovers acetylene (ethyne). It soon becomes an important gas for fuel and lighting.

**1838** German astronomer and mathematician Friedrich Bessel (1784–1846) for the first time determines the distance between the Earth and a star (other than the Sun), finding that 61-Cygni is about 11.4 light-years from Earth.

**1839** French painter and physicist Louis Daguerre (1787–1851) takes the first photograph of the Moon.

The Ages of Steam and Electricity

## Technology

**1839** American inventor Charles Goodyear (1800–60) develops the process of vulcanizing rubber to harden it (by heating raw rubber with sulfur).

**1841** English engineer Joseph Whitworth (1803–87) introduces a system of standard screw threads.

**1841** English inventor Charles Gregory (1817–98) patents a semaphore signal for railroads.

**1841** German-born American engineer John Roebling (1806–69) invents a machine for making wire rope (cable). It becomes important in constructing suspension bridges.

**1841** English chemist Alexander Parkes (1813–90) invents a process for cold vulcanizing rubber (using carbon disulfide and no heat).

**1842** The first suspension bridge in the U.S. opens in Fairmont, Philadelphia, designed by Charles Ellet (1810–62).

**1843** English engineer Thomas Crampton (1816–88) invents the Crampton railroad locomotive. It has a single pair of large driving wheels located behind the firebox and cylinders halfway along the boiler.

**1844** American inventor Samuel Morse (1791–1872) sends the first message on a telegraph line in the U.S. (Washington to Baltimore).

## Biology and Medicine

**1840** French agricultural chemist Jean Boussingault (1802–87) recognizes that nitrates in the soil act as a source of nitrogen for plants.

**1842** English zoologist Richard Owen (1804–92) coins the word "dinosaur" (meaning "terrible lizard" in Greek) to describe certain prehistoric reptiles.

**1842** American surgeon Crawford Long (1815–78) uses ether as an anesthetic.

**1844** American dentist Horace Wells (1815–48) introduces nitrous oxide ("laughing gas") as an anesthetic.

## Physical Sciences and Math

**1840** German chemist Hermann von Fehling (1812–85) devises a test for sugars (called Fehling's test) that relies on the reducing action of sugar on a solution of a copper salt.

**1842** German physicist Julius von Mayer (1814–78) first states the principle of conservation of energy (that energy can be neither created nor destroyed, merely changed from one form into another form).

**1842** Austrian physicist Christian Doppler (1803–53) describes the Doppler effect—the change in frequency (or wavelength) of a wave motion when the source and the observer move closer to or farther away from each other.

**1842** German physicist Julius von Mayer (1814–78) measures the mechanical equivalent of heat, the amount of mechanical work that can be done by one unit of thermal energy. This is determined independently by English physicist James Joule (1818–89) in 1843 and French physicist Gustave Hirn (1815–90) in 1847.

**1843** German astronomer Heinrich Schwabe (1789–1875) declares that sunspots have a cycle of about 10 years (they reach a maximum number every 10 years).

# Timelines 1845–1854

## Technology

**1845** Scottish engineer Robert Thomson (1822–73) patents a pneumatic tire made of vulcanized rubber for horse-drawn vehicles.

**1845** Designed by engineer Isambard Kingdom Brunel (1806–59), SS *Great Britain*, the first successful propeller-driven ship, makes its maiden voyage.

**1845** The first clipper ship, the *Rainbow*, is launched in the U.S.

**1846** American industrialist John Deere (1804–86) markets a plow with a steel moldboard.

**1848** English inventor John Stringfellow (1799–1883) builds a model steam-driven airplane.

**1848** English engineer William Adams (1797–1872) constructs a steam-powered railcar.

**1848** American inventor James Bogardus (1800–74) introduces a method of making cast-iron buildings.

**1849** French gardener Joseph Monier (1823–1906) invents reinforced concrete (concrete cast around iron or steel reinforcing rods).

**c.1850** German chemist Robert Bunsen (1811–99) begins using the Bunsen burner, which was probably designed by his assistant Peter Desaga.

## Biology and Medicine

**1845** English clergyman and amateur biologist Miles Berkeley (1803–89) shows that the disease potato blight is caused by a fungus. By 1848, the potato famine has killed more than one million people in Ireland.

**1847** French physiologist Pierre Flourens (1794–1867) first uses chloroform (trichloromethane) as an anesthetic on small animals. In the same year, Scottish obstetrician James Simpson (1811–70) uses it on humans.

**1850** German scientist Hermann von Helmholtz (1821–94) measures the speed of a nerve impulse.

## Physical Sciences and Math

**1845** Two French physicists—Armand Fizeau (1819–96) and Léon Foucault (1819–68)—take detailed photographs of the Sun.

**1846** German astronomer Johann Galle (1812–1910) is the first person to observe the planet Neptune. A few months later, English astronomer William Lassell (1799–1880) discovers Triton, Neptune's moon.

**1846** The Smithsonian Institution is founded in Washington, D.C., financed with money left by English chemist James Smithson (1765–1829). Its first director is American physicist Joseph Henry (1797–1878).

**1846** Italian chemist Ascanio Sobrero (1812–88) discovers nitroglycerin (trinitroglycerol), a powerful but unstable explosive, which is later found to be useful in the treatment of certain heart disorders.

**1847** German scientist Hermann von Helmholtz (1821–94) and English physicist James Joule (1818–89) establish the principle of the conservation of energy, first proposed in 1842 by German physicist Julius von Mayer (1814–78).

**1848** Scottish physicist William Thomson, Lord Kelvin (1824–1907), introduces the absolute temperature scale (also known as the Kelvin scale). Temperatures on this scale are now designated in units called kelvin, the SI unit of temperature (e.g., absolute zero is 0 K).

## Technology

**1850** American engineer Charles Page (1812–68) builds the first electric locomotive in the U.S.

**1850** A tubular railroad bridge is opened across the Menai Strait, North Wales, designed by English engineer Robert Stephenson (1803–59).

**1851** American inventor Isaac Singer (1811–75) patents a single-thread sewing machine that produces continuous and curved stitching.

**1851** The Crystal Palace, a cast-iron and glass building designed by English architect Joseph Paxton (1801–65), is put up in Hyde Park, London, for the Great Exhibition.

**1852** French engineer Henri Giffard (1825–82) builds a steerable, hydrogen-filled nonrigid airship called a dirigible.

**1852** American inventor Elisha Otis (1811–61) patents the safety elevator; the first is installed in Yonkers, NY, in 1857.

**1853** English inventor George Cayley (1773–1857) constructs a glider capable of carrying a person.

**1854** American gunsmiths Horace Smith (1808–93) and Daniel Wesson (1825–1906) produce a single-action cartridge revolver (patented 1856).

## Biology and Medicine

**1850** English physician Alfred Higginson (1808–84) invents the hypodermic needle. It is used initially for extracting samples of blood or other fluid.

**1853** The world's largest tree, a giant sequoia, is found in California and named *Wellingtonia gigantea* (now *Sequoiadendron giganteum*).

**1854** English physician John Snow (1813–58) pinpoints a link between cholera and contaminated drinking water during an epidemic in London.

**1854** Dutch physician Antonius Mathijsen (1805–78) introduces the plaster cast to hold a broken limb immobile while it heals.

## Physical Sciences and Math

**1851** English astronomer William Lassell (1799–1880) discovers Ariel and Umbriel, two moons of Uranus.

**1852** English chemist Edward Frankland (1825–99) introduces the concept of valence: the combining power of an atom, defined as the number of atoms of hydrogen (or its equivalent) that it combines with.

**1852** English physicist James Joule (1818–89) and Scottish physicist William Thomson, Lord Kelvin (1824–1907), discover the Joule–Thomson effect. This phenomenon is the fall in temperature that occurs when a gas expands through a narrow opening into a region of lower pressure (the principle of the modern refrigerator).

**1853** French chemist Charles Gerhardt (1816–56) derives acetylsalicylic acid (the basis of aspirin) from plants.

**1854** German mathematician Bernhard Riemann (1826–66) develops a type of non-Euclidean geometry later to find applications in relativity theory.

**1854** English astronomer George Airy (1801–92) determines the mass of the Earth from measurements of gravity.

**1854** American chemist David Alter (1807–81) studies atomic spectra, the light given off by elements when heated to incandescence, and uses them as a method of chemical analysis.

# Timelines 1855–1864

## Technology

**1855** Swedish chemist Johan Lundström (1815–88) patents the safety match.

**1855** Italian physicist Luigi Palmieri (1807–96) designs a seismograph, an instrument that detects and measures the strength of earthquakes.

**1856** English steelmaker Henry Bessemer (1813–98) develops the Bessemer converter for making steel out of iron.

**1857** English inventor Edward Cowper (1819–93) creates the hot-blast stove for blast furnaces and improves the steelmaking process.

**1858** The SS *Great Eastern*, designed by English engineer Isambard Kingdom Brunel (1806–59), is launched in England. At the time it is the world's largest ship.

**1859** The first ironclad (wooden-hulled) warship, *La Gloire*, is built in France.

**1859** Belgian engineer Étienne Lenoir (1822–1900) devises a gas-burning internal combustion engine.

**1859** American oil pioneer Edwin Drake (1819–80) drills the world's first productive oil well, in Pennsylvania.

**1859** French chemist Gaston Planté (1834–89) produces the lead–acid accumulator. It becomes the battery used regularly in automobiles.

## Biology and Medicine

**1855** German pathologist Rudolph Virchow (1821–1902) observes that cells originate from the division of other cells.

**1855** French physiologist Claude Bernard (1813–78) establishes that the arrow-poison drug curare can be used as a muscle relaxant.

**1857** French chemist Louis Pasteur (1822–95) observes that microorganisms cause fermentation.

**1859** English naturalist Charles Darwin (1809–82) publishes *On the Origin of Species*, the book in which he puts forward his theory of evolution.

## Physical Sciences and Math

**1855** English chemist Alexander Parkes (1813–90) develops celluloid (patented 1856).

**1856** Scottish physicist William Thomson, Lord Kelvin (1824–1907), coins the term "kinetic energy" to describe energy associated with movement.

**1857** French physicist Léon Foucault (1819–68) devises a method of silver-coating glass for use as mirrors in telescopes.

**1858** German chemist Friedrich Kekulé (1829–96) determines that the normal valence of carbon is four. His findings form the groundwork for theoretical organic chemistry.

**1858** Italian chemist Stanislao Cannizzaro (1826–1910) distinguishes between atomic weights and molecular weights, thereby establishing atomic and molecular weights as the basis of chemical calculations.

**1859** English astronomer Richard Carrington (1826–75) observes flares in the chromosphere (outer layer) of the Sun.

**1860** German chemist Johann Greiss (1829–88) identifies diazo compounds (organic compounds containing nitrogen). He later develops synthetic dyes called azo dyes.

## Technology

**1860** German botanist Julius von Sachs (1832–97) introduces hydroponics, a method of growing plants without soil but using water and chemical fertilizers.

**1860** The Cabin John Bridge, the world's longest masonry arch, is completed. It is designed by American engineer Montgomery Meigs (1816–92).

**1861** Scottish-born Australian James Harrison (1816–93) constructs the world's first large-scale meat-freezing plant, in which meat is prepared for export to Britain.

**1863** French chemist Louis Pasteur (1822–95) introduces pasteurization, a process that kills bacteria. At first it is used for treating wine.

**1863** The world's first underground railroad, the Metropolitan Line, opens in London using steam-hauled trains.

**1863** Built by French engineer Simon Bourgeois, *Le Plongeur*, a 140-foot (42.6-m) compressed air-driven submarine, is launched in France.

**1864** French engineer Pierre Michaux (1813–83) makes the first pedal bicycle.

**1864** American engineer William Sellers (1824–1905) introduces a system of standard screw threads in the U.S.

## Biology and Medicine

**1860** The first fossil of *Archaeopteryx* (a prehistoric flying reptile with some birdlike features) is found in a German quarry.

**1860** English nurse Florence Nightingale (1820–1910) establishes the world's first training school for nurses in London.

**1862** German physician Felix Hoppe-Seyler (1825–95) establishes the presence of hemoglobin, the red oxygen-carrying pigment in the blood.

**1862** German botanist Julius von Sachs (1832–97) establishes that starch is produced in plants by photosynthesis.

## Physical Sciences and Math

**1861** English physicist William Crookes (1832–1919) discovers the element thallium, using Robert Bunsen's new spectrographic technique.

**1862** Belgian chemist Ernest Solvay (1838–1922) patents an industrial process for making soda (sodium carbonate) from chalk (calcium carbonate) and salt (sodium chloride). It is called the Solvay process or ammonia–soda process.

**1862** American astronomers Lewis Swift (1820–1913) and Horace Tuttle (1837–1923) discover comet Swift–Tuttle, which is responsible for the annual Perseid meteor shower.

**1863** Italian Jesuit priest and astronomer Angelo Secchi (1818–78) begins a four-year study of the stars, leading to a system of classifying stars according to their spectral type.

**1864** Scottish physicist James Clerk Maxwell (1831–79) publishes Maxwell's equations, which mathematically describe various electromagnetic phenomena.

**1864** Norwegian chemists Cato Guldberg (1836–1902) and Peter Waage (1833–1900) determine the law of mass action. It states that the rate of a chemical reaction is proportional to the product of the concentrations (active masses) of the reactants.

# Timelines 1865–1876

## Technology

**1865** American locksmith Linus Yale (1821–68) perfects the cylinder lock, and receives a second patent for it.

**1866** English engineer Robert Whitehead (1823–1905) builds a self-propelled torpedo, powered by compressed air.

**1867** Swedish chemist Alfred Nobel (1833–96) patents dynamite in Britain. (U.S. patent 1868).

**1867** English-born American engineer Andrew Hallidie (1836–1900) invents the cable car, a type of streetcar that is hauled by a cable beneath the road surface.

**1868** American engineer George Westinghouse (1846–1914) designs the air brake for trains.

**1869** The Union Pacific transcontinental railroad is completed.

**1869** The Suez Canal is completed by French engineer Ferdinand de Lesseps (1805–94).

**1869** American engineer Thomas Edison (1847–1931) makes an improved ticker-tape machine.

**1871** English inventor James Starley (1830–81) patents his "ordinary" bicycle with one large and one small wheel. It is nicknamed "penny-farthing" (a penny was a large coin and a farthing was a small coin).

## Biology and Medicine

**1865** Austrian monk Gregor Mendel (1822–84) formulates Mendel's laws of inheritance.

**1865** German botanist Julius von Sachs (1832–97) identifies chloroplasts, the structures in green leaves that contain chlorophyll.

**1865** French physician Jean Villemin (1827–92) discovers tuberculosis is infectious.

**1867** English surgeon Joseph Lister (1827–1912) introduces phenol (carbolic acid) as a disinfectant in the operating room.

**1869** Swiss pathologist Johann Miescher (1844–95) isolates deoxyribonucleic acid (DNA), which he calls "nuclein."

## Physical Sciences and Math

**1865** Belgian chemist Jean Servais Stas (1813–91) devises the first modern table of atomic weights, using the element oxygen as a standard (set at 16).

**1866** English astronomer William Huggins (1824–1910) and English chemist William Miller (1817–70) discover the gaseous nature of some nebulas by studying their spectra.

**1866** Italian astronomer Giovanni Schiaparelli (1835–1910) demonstrates that meteor showers are associated with the orbits of comets.

**1868** English astronomer William Huggins (1824–1910) detects a Doppler shift (a shift to a longer wavelength) in the spectrum of the star Sirius, and thereby demonstrates that the star is receding from Earth.

**1869** Russian chemist Dmitri Mendeleev (1834–1907) compiles the first Periodic Table of the Elements.

**1871** Irish physicist George Stoney (1826–1911) observes that three lines in the hydrogen spectrum have wavelengths in a simple ratio to each other. This finding is to be significant in interpreting atomic structure.

## Technology

**1872** American gunsmith Benjamin Hotchkiss (1826–1885) invents a revolving-barrel machine gun.

**1873** French engineer Amédée Bollée (1844–1917) constructs a steam-powered car.

**1874** The three-arch St. Louis Mississippi Bridge is built, designed by American engineer James Eads (1820–87).

**1875** Steam trains begin to run on an elevated railroad in New York City.

**1875** A safe loading level for ships, known as the Plimsoll line, is introduced by an Act of the British Parliament. The Plimsoll line becomes internationally recognized.

**1876** Scottish-born American engineer Alexander Graham Bell (1847–1922) patents the telephone.

**1876** American librarian Melvil Dewey (1851–1931) introduces the Dewey Decimal Classification system for cataloging library books.

**1876** American engineer Melville Bissell (1843–89) invents a carpet sweeper.

**1876** German engineer Nikolaus Otto (1832–91) builds a four-stroke internal combustion engine fueled by coal gas.

## Biology and Medicine

**1872** English-born chemist Robert Chesebrough (1837–1933) patents the process of making petroleum jelly. It is sold as a soothing ointment under the name of Vaseline.

**1873** Austrian physician Josef Breuer (1842–1925) discovers the sensory function of the semicircular canals in the ear.

**1873** Canadian physician William Osler (1849–1919) discovers blood platelets (thrombocytes), small cells that are important in blood clotting.

**1875** German physicians Wilhelm Erb (1840–1921) and Carl Westphal (1833–90) discover the knee-jerk reflex.

## Physical Sciences and Math

**1872** German mathematician Richard Dedekind (1831–1916) publishes his theory of irrational numbers.

**1872** German chemist Eugen Baumann (1846–96) prepares PVC (polyvinyl chloride). Much later, it becomes an important plastic.

**1873** English astronomer Richard Proctor (1837–88) proposes that craters on the Moon were caused by the impact of meteorites, not by volcanoes as previously thought.

**1873** Scottish physicist James Clerk Maxwell (1831–79) publishes *Electricity and Magnetism*, containing his electromagnetic theory of light—that light is a form of electromagnetic radiation (as also are radio waves and X-rays).

**1874** German chemist Othmar Zeidler (1859–1911) synthesizes DDT (dichlorodiphenyltrichloroethane) but does not realize its value as an insecticide.

**1876** German physicist Eugen Goldstein (1850–1930) uses the name "cathode rays" to describe a stream of rays/particles emitted by the cathode in a discharge tube.

# Timelines 1877–1890

## Technology

**1877** English physicist Joseph Swan (1828–1914) and American Thomas Edison (1847–1931) develop the first electric light bulbs.

**1877** American engineer Thomas Edison (1847–1931) and Frenchman Charles Cros (1842–88) independently invent the phonograph.

**1877** German aeronautical pioneer Otto Lilienthal (1848–96) builds his first model glider.

**1878** Scottish engineer Dugald Clerk (1854–1932) builds a two-stroke engine (patented 1881).

**1878** English-born American inventor David Hughes (1831–1900) devises the carbon microphone.

**1879** English inventor Henry Lawson invents the "safety" bicycle, driven by a chain to the rear wheel. The design soon becomes adopted worldwide.

**1882** The first U.S. steam-powered electricity-generating plant, designed by American engineer Thomas Edison (1847–1931), opens on Pearl Street, New York City.

**1883** Belgian engineer Étienne Lenoir (1822–1900) invents the spark plug (for internal combustion engines).

**1884** American-born English inventor Hiram Maxim (1840–1916) makes the first fully automatic machine gun.

## Biology and Medicine

**1877** German bacteriologist Robert Koch (1843–1910) develops a method of staining bacteria (in order to study them through a microscope).

**1880** French parasitologist Alphonse Laveran (1845–1922) discovers the microorganism (a plasmodium) that causes malaria.

**1880** German bacteriologist Karl Eberth (1835–1926) discovers the bacterium that causes typhoid fever.

**1882** German bacteriologist Robert Koch (1843–1910) discovers the bacterium that causes tuberculosis. A year later he discovers the bacterium that causes cholera.

## Physical Sciences and Math

**1877** American astronomer Asaph Hall (1829–1907) discovers Phobos and Deimos, the two moons of Mars.

**1879** American chemists Ira Remsen (1846–1927) and Constantine Fahlberg (1850–1910) discover saccharin, an artificial sweetener 2,000 times sweeter than sugar.

**1880** Scottish astronomer George Forbes (1849–1936) predicts the existence of "Planet X" orbiting beyond Uranus. Pluto was found in 1930.

**1881** German scientist Hermann von Helmholtz (1821–94) demonstrates that hydrogen atoms have their electric charges in whole-number portions, implying that there is a unit of electrical charge (the charge has a finite minimum value).

**1881** English mathematician John Venn (1834–1923) publishes *Symbolic Logic*, in which he introduces his ideas on logical relationships and develops Venn diagrams.

**1882** American physicist Albert Michelson (1852–1931) publishes his first calculation of the speed of light as 186,320 miles per second (299,853 km/s); it is within 0.02 percent of the correct value.

## Technology

**1885** German engineer Karl Benz (1844–1929) builds a three-wheeled car.

**1885** A steel-framed building in Chicago, the world's first skyscraper, is completed.

**1885** German chemist Karl Auer von Welsbach (1858–1929) patents the gas mantle, which produces an incandescent white light in a coal-gas flame.

**1886** American chemist John Pemberton (1831–88) invents Coca-Cola.

**1886** German engineer Gottlieb Daimler (1834–1900) produces a four-wheeled gasoline-engined car.

**1887** Scottish inventor John Dunlop (1840–1921) develops the pneumatic tire for bicycles.

**1887** German-born American engineer Emile Berliner (1851–1929) devises the disk gramophone record.

**1889** The Eiffel Tower is completed in Paris, France, designed by French engineer Gustave Eiffel (1832–1923).

**1889** American undertaker Almon Strowger (1839–1902) makes the first automatic telephone exchange (patented 1891).

**1890** English engineer Herbert Akroyd Stuart (1864–1927) invents a compression-ignition internal combustion engine.

## Biology and Medicine

**1884** Czech-born American surgeon Carl Koller (1857–1944) introduces the use of cocaine as a local anesthetic.

**1885** Austrian neurologist Sigmund Freud (1856–1939) develops psychoanalysis as a diagnostic procedure.

**1885** French chemist Louis Pasteur (1822–95) produces a vaccine against rabies.

**1890** Russian physiologist Ivan Pavlov (1849–1936) begins the experiments that lead to the discovery that digestive stomach secretions are stimulated by nerve impulses.

## Physical Sciences and Math

**1885** A supernova flares up in the Andromeda galaxy; no other supernova visible to the naked eye will appear until 1987.

**1886** German chemist Clemens Winkler (1838–1904) discovers germanium, whose existence was predicted in 1869 by Dmitri Mendeleev (1834–1907).

**1887** German chemist Emil Fischer (1852–1919) synthesizes fructose, the sugar found in fruits.

**1888** German physicist Heinrich Hertz (1857–94) detects radio waves.

**1889** American astronomer Edward Barnard (1857–1923) takes the first photographs of the Milky Way (our galaxy).

**1889** English chemist Frederick Abel (1827–1902) and Scottish physicist James Dewar (1842–1923) develop the propellant explosive cordite.

**1889** Irish physicist George Fitzgerald (1851–1901) develops a theory (the Fitzgerald contraction) that at speeds approaching the speed of light, an object's length decreases. It is later used by German–American physicist Albert Einstein (1879–1955) to formulate his theory of relativity.

Timelines

# Timelines 1891–1900

## Technology

**1891** American inventor George Blickensderfer (1850–1917) patents a portable typewriter.

**1891** German aeronautical pioneer Otto Lilienthal (1848–96) makes a steerable human-carrying glider.

**1892** French engineer François Hennebique (1842–1921) develops prestressed concrete, in which the reinforcing rods are tensioned before the concrete is poured.

**1892** Scottish physicist James Dewar (1842–1923) invents the vacuum bottle (sold as a flask for domestic use under the trade name Thermos in 1904).

**1892** A telephone line linking New York City and Chicago goes into operation.

**1893** German physicists Julius Elster (1854–1920) and Hans Geitel (1855–1923) make the first practical photoelectric cell.

**1893** English-born Australian inventor Lawrence Hargrave (1850–1915) invents a human-carrying box kite.

**1893** American engineer Whitcomb Judson (d. 1905) patents the "clasp locker," a forerunner of the zipper, designed for fastening shoes.

## Biology and Medicine

**1891** Dutch anthropologist Eugène Dubois (1858–1940) discovers fossils of the human ancestor Java man, which he names *Pithecanthropus*. Today it is classified as *Homo erectus*.

**1893** African–American surgeon Daniel Williams (1858–1931) performs the first open-heart surgery operation.

**1895** Australian-born Scottish physician David Bruce (1855–1931) shows that the tsetse fly carries the parasite *Trypanosoma brucei*, which causes sleeping sickness.

**1895** English botanist Frederick Blackman (1866–1947) discovers that gas exchange (oxygen and carbon dioxide) in plants takes place through pores in the leaves.

## Physical Sciences and Math

**1891** Irish physicist George Stoney (1826–1911) coins the term "electron" for the fundamental unit of electricity (from his earlier "electrine" of 1874).

**1892** English organic chemist William Tilden (1842–1926) makes synthetic rubber from isoprene.

**1894** Scottish physicist James Dewar (1842–1923) makes liquid oxygen, one of the first elemental gases to be liquefied.

**1894** English physicist John Strutt, Lord Rayleigh (1842–1919), and Scottish chemist William Ramsay (1852–1916), discover the inert gas argon in air.

**1895** French physicist Jean Perrin (1870–1942) establishes that cathode rays consist of particles carrying a negative charge. They are, in fact, electrons.

**1895** German physicist Wilhelm Röntgen (1845–1923) takes the first X-ray.

**1896** French physicist Henri Becquerel (1852–1908) discovers radioactivity.

**1896** German biochemist Albrecht Kossel (1853–1927) discovers the amino acid histidine, which is essential in the diet of animals.

## Technology

**1893** German inventor Rudolf Diesel (1858–1913) first builds a compression-ignition engine (diesel engine), patented in 1892; a fully operational engine was demonstrated in 1897.

**1893** French brothers Auguste (1862–1954) and Louis (1864–1948) Lumière invent a motion-picture camera.

**1895** American inventor King Gillette (1855–1932) first has the idea for the disposable double-edged razor blade.

**1896** American engineer and patent lawyer Charles Curtis (1860–1953) invents an impulse steam turbine, later used in ships and electricity power plants.

**1897** German physicist Ferdinand Braun (1850–1918) develops the cathode-ray tube, later to be central to radar and television sets.

**1898** German gunmaker George Luger produces the Luger automatic pistol.

**1898** Irish–American schoolmaster and engineer John Holland (1840–1914) builds the first modern submarine.

**1900** The rigid airship *LZ-1*, designed by German engineer Graf Ferdinand von Zeppelin (1838–1917), makes its maiden flight.

**1900** American engineer Thomas Edison (1847–1931) invents the nickel–iron accumulator (Ni–Fe cell).

## Biology and Medicine

**1896** English bacteriologist Almroth Wright (1861–1947) develops a vaccine against typhoid fever.

**1898** English physiologist John Langley (1852–1925) distinguishes the autonomic nervous system, the part of the nervous system that controls the involuntary muscles and glands.

**1900** Austrian-born American pathologist Karl Landsteiner (1868–1943) discovers A, B, and O blood groups and their intercompatability.

**1900** Austrian neurologist Sigmund Freud (1856–1939) publishes his seminal book *The Interpretation of Dreams*.

## Physical Sciences and Math

**1896** Dutch physicist Pieter Zeeman (1865–1943) observes the Zeeman effect, which is the splitting of atomic spectral lines in a strong magnetic field.

**1896** New Zealand-born British physicist Ernest Rutherford (1871–1937) describes and names alpha particles (which are helium nuclei) and beta particles (which are electrons).

**1896** Swedish physical chemist Svante Arrhenius (1859–1927) proposes a link between the levels of carbon dioxide ($CO_2$) in the atmosphere and global temperatures.

**1897** English physicist J.J. Thomson (1856–1940) identifies the electron, the first subatomic particle.

**1898** French physical chemists Marie (1867–1934) and Pierre (1859–1906) Curie discover the elements polonium and radium.

**1900** German physicist Paul Drude (1863–1906) reveals that electrons carry electricity in an electric current.

**1900** French physicist Henri Becquerel (1852–1908) shows that beta particles are, in fact, electrons.

**1900** German physicist Max Planck (1858–1947) analyzes black-body radiation and proposes the quantum theory: that radiation is emitted in separate "packets," or quanta.

# Timelines 1901–1910

## Technology

**1901** Italian physicist and radio pioneer Guglielmo Marconi (1874–1937) makes the first transatlantic radio transmission.

**1901** German physicist Ferdinand Braun (1850–1918) invents the crystal detector for tuning a radio.

**1903** American brothers Wilbur (1867–1912) and Orville Wright (1871–1948) make the first sustained flight in a heavier-than-air airplane.

**1903** Dutch physiologist Willem Einthoven (1860–1927) invents the electrocardiograph for recording the electrical activity of the heart.

**1904** The first section of the New York City subway opens, with electric trains serving 28 stations.

**1904** English engineer John Fleming (1849–1945) invents the diode valve (vacuum tube).

**1906** American engineer Lee De Forest (1873–1961) invents the triode valve (vacuum tube, or audion tube).

**1906** Canadian-born American electrical engineer Reginald Fessenden (1866–1932) makes AM (amplitude modulation) radio transmissions.

**1906** Russian physicist Boris Golitsyn (1862–1916) creates an electromagnetic seismograph for detecting earthquakes.

## Biology and Medicine

**1903** German surgeon Georg Perthes (1869–1927) first uses X-rays to treat cancerous tumors.

**1903** Russian physiologist Ivan Pavlov (1849–1936) develops the concept of the conditioned reflex (the way in which an action can be influenced by previous repetitive behavior).

**1906** English biochemist Frederick Gowland Hopkins (1861–1947) concludes that "accessory food factors," later called vitamins, are essential to health.

**1906** Austrian physician Clemens von Pirquet (1874–1929) shows that hay fever is an allergic reaction to pollen, and devises the term "allergy."

## Physical Sciences and Math

**1902** German chemist Emil Fischer (1852–1919) determines that proteins are polypeptides (i.e., made up of chains of amino acids).

**1902** French meteorologist Léon Teisserenc de Bort (1855–1913) distinguishes the stratosphere and troposphere layers in the Earth's atmosphere.

**1902** French chemist Georges Claude (1870–1960) develops a process for the large-scale liquefaction of air.

**1904** English physicist J.J. Thomson (1856–1940) puts forward his model of the atom—a spherical mass of positively charged matter with electrons embedded in it.

**1905** American amateur astronomer Percival Lowell (1855–1916) takes the first photograph of Mars.

**1905** German-born American physicist Albert Einstein (1879–1955) publishes his special theory of relativity.

**1906** New Zealand-born English physicist Ernest Rutherford (1871–1937) measures the ratio of charge to mass for alpha particles and deduces that they are helium nuclei.

**1906** French mathematician Maurice Fréchet (1878–1973) introduces functional calculus to mathematics.

## Technology

**1906** French physicist Jacques d'Arsonval (1851–1940) invents a freeze-drying technique used to preserve foods.

**1907** Belgian-born American chemist Leo Baekeland (1863–1944) invents Bakelite plastic.

**1907** French engineer Louis Bréguet (1880–1955) constructs a primitive kind of helicopter, the gyroplane.

**1907** The German company Henkel markets the first synthetic household detergent (Persil).

**1908** The first Model T Ford from American industrialist Henry Ford (1863–1947) comes off the assembly line at Detroit.

**1908** The American Corning Flint Glassworks funds a research project to develop heat-resistant glass for railroad lanterns. (It leads to the invention of Pyrex cookware in 1915.)

**1909** French aviator Louis Blériot (1872–1936) flies across the English Channel.

**1909** "SOS" is introduced as the international radio distress signal.

**1910** French chemist Georges Claude (1870–1960) invents neon discharge tubes for use in lighting and signs.

**1910** The Manhattan Bridge is completed in New York City.

## Biology and Medicine

**1906** English biologist William Bateson (1861–1926) coins the term "genetics."

**1908** French anthropologist Marcellin Boule (1861–1942) reconstructs the first complete skeleton of an early Neanderthal human found in France.

**1909** Danish botanist Wilhelm Johannsen (1857–1927) coins the term "gene" for the factor that carries inheritable characteristics.

**1910** American pathologist Francis Rous (1879–1970) identifies the first cancer-causing virus.

## Physical Sciences and Math

**1906** Irish-born English geologist Richard Oldham (1858–1936) deduces that the Earth's core is molten.

**1907** Swiss chemist Jacques Brandenburger (c.1873–1954) prepares cellophane (patent 1908; trademark 1912).

**1908** Danish astronomer Ejnar Hertzsprung (1873–1967) introduces a method of classifying stars by plotting a graph of luminosity against temperature. It becomes the basis of the Hertzsprung–Russell diagram, published in 1913.

**1908** American physicist William Coolidge (1873–1975) uses tungsten to make an incandescent filament lamp.

**1909** Danish chemist Søren Sørensen (1868–1939) introduces the concept of pH, a measure of hydrogen ion concentration, and therefore of the strength of an acid or alkali.

**1909** American Edward Davidson first uses carbon tetrachloride (tetrachloromethane) as a fire extinguisher.

**1910** New Zealand-born English physicist Ernest Rutherford (1871–1937) proves the existence of the nucleus of the atom (announced in 1911) after seeing an experiment by his assistant Ernest Marsden (1889–1970).

# Glossary

**aerodynamic** Designed to reduce or minimize the resistance to air flow.

**alloy** Any material made of a blend of a metal with another substance to give it special qualities, such as resistance to corrosion or greater hardness.

**anode** A positively charged plate or electrode.

**coal gas** Flammable gas obtained in the destructive distillation of soft (bituminous) coal, often as a byproduct in the preparation of coke.

**compressed air** Air under greater pressure than the air in the environment, especially when used to power a mechanical device, or to provide a portable supply of oxygen.

**diesel fuel** A fuel composed of distillates obtained in the petroleum refining process; it has an ignition temperature of 540ºC (1,000ºF) and is ignited in engines by the heat generated from compressed air.

**electrical conductor** Any substance that allows an electric current to flow through it.

**electronics** A branch of physics and of electrical engineering that involves the manipulation of voltages and electric currents through the use of various devices.

**gasoline fuel** A volatile, flammable liquid composed of distillates obtained in the petroleum refining process with a boiling range of 30ºC to 200ºC (85ºF to 390ºF) and ignited in engines by a spark.

**gene** The basic unit of inheritance that controls a characteristic of an organism.

**glider** A light, unpowered aircraft designed to glide after being towed into the air or launched from a catapult.

**helium** A light, colorless, nonflammable inert gaseous element (atomic number 2) occurring in natural gas, in radioactive ores, and in small amounts in the atmosphere; chemical symbol He.

**hydrogen** A colorless, highly flammable gaseous element (atomic number 1), the lightest of all gases, and the most abundant element in the universe; symbol H.

**monoplane** An airplane with only one pair of wings.

**Morse code** A telegraph code in which letters and numbers are represented by strings of dots and dashes.

**platinum** A silvery-white, malleable, ductile, metallic element often used as a chemical catalyst; platinum is highly resistant to corrosion and tarnish.

**semaphore** A method or device used for the visible transmission of messages, using lights, flags, or pivoted arms.

**SI units** The International System of Units (abbreviated SI from the French *Systeme Internationale*) is a scientific method of expressing the magnitudes or quantities of important natural phenomena, such as length, time, and electrical current.

**speed of sound** The speed at which sound travels in a given medium under specified conditions; at sea level, it is 760 miles per hour (1,220 km/h).

**static** A hissing or crackling noise caused by electrical interference from the atmosphere.

**vaccine** A preparation containing viruses or other microorganisms, introduced into the body to stimulate the formation of antibodies and build up immunity.

**vacuum tube** An airtight glass tube in which electricity is conducted by electrons passing through a partial vacuum from a cathode to an anode; also called an electron tube.

**wrought iron** A form of iron that is tough, malleable, and relatively soft; it contains usually less than 0.1 percent carbon and 1 or 2 percent of slag.

# Index

Words and page numbers in **bold** type indicate main references to the various topics. Page numbers in *italic* refer to illustrations. An asterisk before a page range indicates mentions on each page rather than unbroken discussion

## A

Abel, Frederick 101
accumulators 54, *55*, 96, 103
acetylene 92
Adams, William 94
Ader, Clément 78
**airplanes 76–81**, *81*, 104
**airships 70–5**, 95, 103
Airy, George 95
*Akron* (airship) 75
Albert, Prince 16
algebra 91
alleles 21–2
allergy 104
Alter, David 95
*Archaeopteryx* 97
arched bridges 97, 99
Armstrong, Edwin 59
Arrhenius, Svante 103
Arsonval, Jacques d' 105
Aston, Francis 65
atomic/molecular weights 96, 98
atomic structure 37–8, *64*
Audubon, John James 90
**automobiles 66–9**, *67*, *68*
  early *49*, *67*
  gasoline-powered 68–9
  Model T 69, *69*, 105
  steam-powered 66–8, 99
  three-wheeled 101
  *see also* internal combustion engine
azo dyes 96

## B

Babbage, Charles 91
bacteria 44–6, *46*
Baden-Powell, Baden 76
Baekeland, Leo Hendrik 88, 105
Baer, Karl von 90
Baily, Francis 42
Baily's beads 42
Bakelite *86–7*, 88, 105
Bang, Bernhard 46
Barnard, Edward 101
Bateson, William 105
batteries 52–5, *53*, *55*
  *see also* accumulators
Bauer, Wilhelm 30
Baumann, Eugen 99
Beaumont, William 91
Becquerel, Alexandre 92
Becquerel, Henri 102, 103
Beijerinck, Martinus 46–7
Bell, Alexander Graham 28, *29*, 82, 99
Benz, Karl 48–9, 68, 69, 101
Berkeley, Miles 94
Berliner, Emile 83, 101
Bernard, Claude 96
Berzelius, Jöns 92
Bessel, Friedrich 92
Bessemer, Henry 12–13, 96
Bessemer converter 13, *14*, *14*, 96
bicycles 96, 98, 100, 101
biology *90–105
Bissell, Melville 99
Blackman, Frederick 102
blast furnaces 11–12, 13, *13*, 96
Blériot, Louis *81*, 105
Blickensderfer, George 102
Blumlein, Alan Dower 84
Bogardus, James 94
Bollée, Amédée 66–8, 99
Boule, Marcellin 105

Bourgeois, Simon 32, 97
Boussingault, Jean 91, 93
Boyden, Uriah 6, 9
Braille, Louis 90
Brandenburger, Jacques 105
Branly, Édouard 56
brass 15
Braun, Ferdinand 103, 104
Brearley, Harold 15
Bréguet, Louis 105
Breuer, Josef 99
bridges 93, 95
  arched 97, 99
Brownian motion 90
Brown, Robert 90, 91
Bruce, David 102
Brunel, Isambard 94, 96
Bunsen, Robert 54, 94
Bunsen burner 94
Burdin, Claude 6
Bushnell, David 30, 34

## C

cadmium cell 54
calculus 104
Cannizzaro, Stanislao 96
carbon tetrachloride 105
Carothers, Wallace Hume 88
carpet sweeper 99
Carrington, Richard 96
cars *see* automobiles
cassette tape 84
Cassini, Giovanni 42–3
cast iron 12
  buildings 94
cathode rays 99, 102, 103
Cayley, George 76, 95
cellophane 105
celluloid 87, 96
Chanute, Octave 78
Chardonnet, Hilaire de 88
Charles, Jacques 70
Chesebrough, Robert 99
chloroform 94

Index 107

chlorophyll 92
cholera 95
Clark, Josiah 54
Clark standard cell 54
Claude, Georges 104, 105
Clerk, Dugald 50, 100
Coca-Cola 101
cocaine 101
Cody, Samuel 76
coke 12
Colt, Samuel 34, 91
concrete 94, 102
conditioned reflex 104
conservation of energy 93, 94
Cooke, William 26, 92
Coolidge, William 105
Cooper, Peter 90
copper 15, 65
cordite 101
Correns, Karl 22
Cort, Henry 12
Cowper, Edward 96
Crampton, Thomas 93
Crookes, William 63, 97
Cros, Charles 100
Crystal Palace
    London *16–17*, 18, 19, 95
    New York 19, *19*
Cugnot, Nicolas-Joseph 66
curare 96
Curie, Marie 103
Curie, Pierre 103
Curtis, Charles 103

# D

Daguerre, Louis 42, 92
Daimler, Gottlieb 49, 68, 101
Daniell, John 52, 92
Daniell cell 52
Darby, Abraham 12
Darwin, Charles 92, 96
Davenport, Thomas 92
Davidson, Edward 105

Davy, Edward 92
Dedekind, Richard 99
Deere, John 94
De Forest, Lee 59–60
de Lesseps, Ferdinand 98
Desaga, Peter 94
detergents 105
*Deutschland* (airship) 74
de Vries, Hugo 22
Dewar, James 101, 102
Dewey, Melvil 99
Diesel, Rudolf 50, 103
diesel engine 49–50, 103
Dietz, Charles 66
Doppler, Christian 93
Doppler effect 93, 98
Drake, Edwin 96
Draper, John 42
Drebbel, Cornelis 30
Drude, Paul 103
Dubois, Eugène 102
Dudgeon, Richard 66
Dunlop, John 101
Duryea, Charles 69
Duryea, Frank 69
Dutrochet, René 90, 92
dynamite 98
dynamo 90

#

Eads, James 99
Eberth, Karl 44, 100
Edison, Thomas 28, 54, 82, 98, 100, 103
Eiffel, Gustave 101
Eiffel Tower 15, 101
Einstein, Albert 101, 104
Einthoven, Willem 104
electric bell 90
**electricity**
    flow 65, *65*
    generating plant 100
    **sources of 52–5**
electrocardiograph 104

electrolysis 91
electromagnetism 56
**electrons 62–5**, 64, *64*, 102, 103
elevator 95
Ellet, Charles 93
Elster, Julius 102
Erb, William 99
Evans, Oliver 66
exhibitions, international 18–19

# F

Fahlberg, Constantine 100
Faraday, Michael 90, 91
Fawcett, Eric 88
Fehling, Hermann von 93
Fessenden, Reginald 58–9, 60, 104
Fischer, Emil 101, 104
Fitzgerald, George 101
Fizeau, Armand 94
Fleming, John 59, 104
Flourens, Pierre 90, 94
*Flyer* airplanes 79, *79*, 80
Forbes, George 100
Ford, Henry 69, 105
Forest, Lee De 59–60, 104
Foucault, Léon 94, 96
Fourier, Joseph 90
Fourneyron, Benoît 6
Fracastoro, Girolamo 44
Francis, James *7*, 9
Francis turbine *7*, *8*
Frankland, Edward 95
Fréchet, Maurice 104
Freud, Sigmund 101, 103
fuel cells 54–5
Fulton, Robert 30

# G

Galilei, Galileo 40–1
Galle, Johann 94

gallium 37
Galvani, Luigi 52
galvanometer 92
Garrett, George 33
gas mantle 101
Gauss, Karl 25–6, 90, 91
Geissler, Heinrich 62
Geissler tubes 62–3, *63*
Geitel, Hans 102
gene 105
**genetics 20–3**, 105
    Mendel's laws 20–1
Gerhardt, Charles 95
germanium 37, 101
**germs 44–7**
Gibson, Reginald 88
Giffard, Henri 70–2, 95
Gillette, King 103
glass
    heat-resistant 105
    silver coating 96
gliders 76–8, 80, 95, 100, 102
Goldstein, Eugen 99
Golitsyn, Boris 104
Goodyear, Charles 93
*Graf Zeppelin* (airship) 74
Graham, Thomas 90
Gram, Hans 44
gramophone 82–3, 101
Gray, Asa 92
Gray, Elisha 28
*Great Britain* (steamship) 94
*Great Eastern* (steamship) 96
**Great Exhibition 16–19**, *16–17*, 95
    exhibitors 18
*Great Western* (steamship) 92
Gregory, Charles 93
Greiss, Johann 96
Grove, William 55
Guldberg, Cato 97
Gurney, Goldsworthy 66

# H

Haenlein, Paul 72
Hall, Asaph 100
Hallidie, Andrew 98
Hargrave, Lawrence 102
Harrison, James 97
helicopter 105
Helmholtz, Hermann von 94, 100
hemoglobin 97
hemophilia 23
Henle, Jacob 44
Hennebique, François 102
Henry, Joseph 25, 26, 90, 91, 94
Henson, William 78
heredity, principles of *21*, 22–3, *23*
Hertz, Heinrich 56, 101
Hertzsprung, Ejnar 105
Hertzsprung-Russell diagram 105
Higginson, Alfred 95
*Hindenburg* (airship) 75
Hipparchus of Nicaea 41
Hirn, Gustave 93
Hittorf, Johann 62
Holland, John 34, 103
*Homo erectus* 102
Hopkins, Frederick Gowland 104
Hoppe-Seyler, Felix 97
Hotchkiss, Benjamin 99
Howd, Samuel 6
Huggins, William 98
Hughes, David 26–7, 100
Hunley, Horace 31
Hyatt, John Wesley 87
hypodermic needle 95

# I

inheritance, Mendel's laws 98

**internal combustion engine 48–51**
    carburetors 48–9
    compression-ignition engine 50, 101
    diesel engine 49–50, 103
    four-stroke 48, 51, *51*, 99
    gas-burning 96
    gasoline engine 48–9
    rotary engines 50–1
    spark plugs 100
    two-stroke 50, *50*, 100
    Wankel engine 51
**iron 10–14**
    puddling process 12
    smelting 10, 90

# J

Jacobi, Moritz von 92
Johannsen, Wilhelm 105
Joule, James 93, 94, 95
Joule-Thomson effect 95
Judson, Whitcomb 102

# K

Kaplan, Viktor 8
Kaplan turbine 8, *8*
Kekulé, Friedrich 96
Kelly, William 13
kinetic energy 96
kites 76, 102
Koch, Robert 44, 46, 100
Koller, Carl 101
Kossel, Albrecht 102
Krebs, Arthur 72

# L

Lake, Simon 34
lamps 105
Landsteiner, Karl 103
Langley, John 103
Langley, Samuel 79–80

Langrenus, Michael 41
Laplace, Pierre-Simon de 91
Lassell, William 94, 95
Laveran, Alphonse 100
Lawson, Henry 100
Le Bris, Jean-Marie 76
Leclanché, Georges 53
Leclanché cell 53
Lecoq de Boisbaudran, Paul 37
Leeuwenhoek, Antonie van 44
Lenoir, Étienne 48, 96, 100
Le Sage, George Louis 25
Lesseps, Ferdinand de 98
Levassor, Émile 68
Liebig, Justus von 91
light 99
  speed of 100, 101
lighting
  electric bulbs 100
  neon tubes 105
Lilienthal, Otto 77–8, 77, 100, 102
Lister, Joseph 98
locomotives, electric 95
Lodge, Oliver 56
Löffler, Friedrich 46
Lohrmann, Wilhelm 42
Long, Crawford 93
loudspeakers 85, 85
Lowell, Francis Cabot 9
Lowell, Percival 43, 104
Lowell, Massachusetts 9
Luger, George 103
Lumière, Auguste 103
Lumière, Louis 103
Lundström, Johan 96
Luppis, Giovanni 35
Lyell, Charles 91

## M

McCormick, Cyrus 91
machine gun 99, 100
*Macon* (airship) 75

magnetophon 84
malaria 100
Manhattan Bridge 105
Maraldi, Giacomo 42
Marconi, Guglielmo 57, *57*, 59, 104
Marcus, Siegfried 48
**Mars** 100, 104
  life on 43
  mapping **42–3**, *43*
Marsden, Ernest 105
mass action 97
mass production 69
mass spectrograph 65
mathematics *90–105
Mathijsen, Antonius 95
Maxim, Hiram 79, 100
Maxwell, James Clerk 56, 97, 99
Maybach, Wilhelm 49
Mayer, Julius von 93, 94
medicine *90–105
Meigs, Montgomery 97
melamine 91
**Mendel, Gregor 20–2**, 98
Mendeleev, Dmitri 36–9, *37*, 98, 101
Michaux, Pierre 97
Michelson, Albert 100
microphones 28–9, 84, 85, *85*, 100
Miescher, Johann 98
Miller, William 98
Monier, Joseph 94
**Moon** 99
  mapping 40–1, **40–2**
  photographs 42, 92
Morse, Samuel 25, 26, 92, 93
Morse Code 26
Murdock, William 66

## N

needle telegraph 91
Neilson, James 90
Neptune 94

New York Crystal Palace 18, 19, *19*
Niépce, Joseph 90
Nightingale, Florence 97
Nilson, Lars 37
nitrocellulose 87
nitroglycerin 94
nitrous oxide 93
Nobel, Alfred 98
Nordenfeldt, Thorsten 33
numbers, irrational 99
nylon 88

## O

Ohm, Georg 90
Ohm's law 90
oil wells 96
Oldham, Richard 105
ornithopter 76
Ørsted, Hans 25, 56
Osler, William 99
osmosis 90
Otis, Elisha 95
Otto, Nikolaus 48, 99
Owen, Richard 93

## P

Page, Charles 95
Paget, James 91
Palmieri, Luigi 96
Panhard, René 68
Parkes, Alexander 87, 93, 96
Parkesine 87
Pasteur, Louis 44, *45*, 96, 97, 101
pasteurization 97
Pavlov, Ivan 101, 104
Paxton, Joseph 17–18, 95
Peacock, George 91
Pelton, Lester 8
Pelton wheel 8, *8*
Pemberton, John 101
**Periodic Table 36–9**, *38–9*, 98

periscopes 34
Perrin, Jean 63, 102
Perthes, Georg 104
petroleum jelly 99
Pfleumer, Fritz 84
phenol 98
phonograph 82, *83*, 100
photoelectric cell 102
photographs 90, 92, 94, 101, 104
Pickard, Greenleaf 59
Pilcher, Percy 76–7
Pirquet, Clemens von 104
Pitman, Isaac 91
Planck, Max 103
Planté, Gaston 54, 96
**plastics 86–9**
Plateau, Joseph 91
Plimsoll line 99
plows 94
Plücker, Julius 62
Pluto 100
polyethylene 88, 89
polymers 89, *89*
polyvinyl chloride (PVC) 99
Poncelet, Jean-Victor 6
Popov, Aleksandr 56
potato blight 94
Poulsen, Valdeman 83
Proctor, Richard 99
psychoanalysis 101

# Q

quantum theory 103

# R

*R 101* (airship) 75
rabies 101
**radio 56–61**, *58*
   amplitude modulation (AM) 59, 104
   broadcasts on 60
   "coherer" 56
   frequency modulation 60

radio (*cont.*)
   heterodyne/superheterodyne circuits 60
   reception 59, 61, *61*
   transistors 59
   transmission 61, *61*, 104
   tuning 60, 104
   vacuum tubes 59
   very high frequency (VHF) 59
   voices on 58–9
radioactivity 102
radiotelegraphy 57, *57*
radiotelephony 58–9
radio waves, detecting 56, 101
railroads 98
   elevated 99
   locomotives 93
Ramsay, William 102
Rathke, Martin 90
rayon 88
razor blades 103
reaping/binding machine 91
Reis, Philipp 28
relativity, theory of 101, 104
relay, electromagnetic 91
Remsen, Ira 100
Renard, Charles 72
revolvers 91, 95
Rickover, Admiral Hyman 34
Riemann, Bernhard 95
Robiquet, Pierre 91
Rochas, Alphonse Beau de 48
Roebling, John 93
Ronalds, Francis 24
Röntgen, Wilhelm 102
Ross, James 91
Rous, Francis 105
rubber
   synthetic 102
   vulcanized 93, 94
Rutherford, Ernest 65, 103, 104, 105

# S

saccharin 100
Sachs, Julius von 97, 98
safety match 96
Salvá i Campillo, Don Francisco 24
scandium 37
Schiaparelli, Giovanni 40, 43, 98
Schilling, Pavel 25
Schmidt, Johann 42
Schütz, Wilhelm 46
Schwabe, Heinrich 93
Schwann, Theodor 92
Schwartz, David 72
screw threads 97
Secchi, Angelo 97
seismograph 96, 104
Sellers, William 97
sewing machine 95
*Shenandoah* (airship) 75
Shiga, Kiyoshi 46
ships, ironclad 96
shorthand 91
Siemens, Friedrich 13
Siemens, William 13
Siemens-Martin furnace 13–14
Simpson, James 94
Singer, Isaac 95
Sirius (star) 98
Sirius (steamship) 92
skyscraper 101
sleeping sickness 102
Smith, Horace 95
Smith, Oberlin 83, 84
Smithson, James 94
Smithsonian Institution 94
Snow, John 95
Sobrero, Ascanio 94
soda 97
Solvay, Ernest 97
Somerville, Mary 91
Sömmering, Samuel von 24
Sørensen, Søren 105

SOS 105
**sound recording 82–5**
  magnetic 83–4
  quality improvement 84
  transducers 85, *85*
stainless steel 15
Stanley, Francis 69
Stanley, Freelan 69
starch 97
Starley, James 98
Stas, Jean Servais 98
steam turbine 103
**steel** 10, **12–15**
Stephenson, George 90
Stephenson, Robert 95
Stoney, George 64, 98, 102
streetcar 92
Stringfellow, John 78, 94
stroboscope 91
Strowger, Almon 101
Strutt, John, Lord Rayleigh 64, 102
Stuart, Herbert Akroyd 50, 101
Sturgeon, William 92
Sturm, Jacques 90
**submarines 30–5**, *31*, *32-3*, 97, 103
Suez Canal 98
sunspots 93
supernovae 101
surgery, open-heart 102
Swan, Joseph 88, 100
Swift, Lewis 97

# T

tape recorder 84
Teisserenc de Bort, Léon 104
**telegraph, electric 24–7**, 25, *27*, 90, 92
  needle telegraph 25–6, *26*
  printing telegraph 27
telegraphone 83
telephone exchange 101

telephone line 102
**telephones** *28*, **28–9**, *29*
  long-distance lines 29
temperature scales 94
thallium 97
thermoplastics 89
Thomson, James 6
Thomson, J.J. 62, 64, 65, 103, 104
Thomson, Robert 94
Thomson, William, Lord Kelvin 94, 95, 96
ticker-tape machine 98
Tilden, William 102
tires 101
Tissandier, Albert 72
Tissandier, Gaston 72
titanium 15
torpedo 34, 98
transistors 59
Trevithick, Richard 66
Tschermak-Seysenegg, Erich von 22
tungsten 15, 65
turbines 6, 7–8
Tuttle, Horace 97
typewriter 102

# V

vaccines 47
vacuum bottle 102
vacuum pump 62
vacuum tubes 104
Vail, Alfred 26
van Leeuwenhoek, Antonie 44
Venn, John 100
Victoria, Queen 17
Villemin, Jean 98
Virchow, Rudolph 96
viruses 46–7, *47*
Volta, Alessandro 24, 52, *53*, 55
voltaic pile 52, *53*
Vries, Hugo de 22

# W

Waage, Peter 97
Wankel, Felix 51
**water turbines 6–9**
Weber, Wilhelm 25–6, 91
Wells, Horace 93
Welsbach, Karl Auer von 101
Wesson, Daniel 95
Western Union Telegraph Company 27
Westinghouse, George 98
Weston, Edward 54
Weston standard cell 54
Westphal, Carl 99
Wheatstone, Charles 26, 90, 92
Whitehead, Robert 34–5, 98
Whitworth, Joseph 93
Williams, Daniel 102
Winkler, Clemens 37, 101
Wright, Almroth 103
Wright, Orville 78, 80–1, 104
Wright, Wilbur 78, *78*, 80–1, 104
Wright *Flyer* 79, *79*
wrought iron 11, 12

# X

X-rays 102, 104

# Y

Yale, Linus 98

# Z

Zédé, Gustave 32
Zeeman, Pieter 103
Zeidler, Othmar 99
Zeppelin, Ferdinand von 73, 103
zeppelin airships 73–5, *73*, 103